Mechanical Behaviour of Materials

Mechanical Behaviour of Materials

Edited by

A. Bakker

Invited papers presented at the

Seventh International Conference on Mechanical Behaviour of Materials - ICM7

The Hague, The Netherlands, May 28 - June 2, 1995

Delft University Press, Delft, The Netherlands

Published and distributed by:

Delft University Press
Stevinweg 1
2628 CN Delft
The Netherlands
Telephone +31 15 783254
Fax +31 15 781661

By order of:

European Structural Integrity Society - ESIS
c/o Materials Laboratory
Delft University of Technology
Rotterdamseweg 137
2628 AL Delft
The Netherlands
Telephone +31 15 785418
Fax +31 15 786730

CIP-DATA KONINKLIJKE BIBLIOTHEEK, DEN HAAG

Bakker, A. (editor)

Mechanical Behaviour of Materials: Invited lectures of the 7th International Conference (ICM-7). - Delft: Delft University Press. - Ill.
- With ref.
ISBN 90-407-1126-7
NUGI 841
Subject headings: Deformation, Damage, Fracture

Copyright © 1995 European Structural Integrity Society

All rights reserved.
No part of the material protected by this copyright notice may be reproduced or utilized in any form or by any means, electronic or mechanical, including photocopying, recording or by any information storage and retrieval system, without permission from the publisher: Delft University Press, Stevinweg 1, 2628 CN Delft, The Netherlands.

Printed in The Netherlands

Contents

Foreword   vii

Effect of Inhomogeneities in the Modelling of Mechanical Behaviour and
    Damage of Metallic Materials   1
       A. Pineau

Micromechanics of Damage in Metals   23
       Viggo Tvergaard

On Nonlinear Constitutive Equations for Elastic and Viscoelastic
    Composites with Growing Damage   45
       R.A. Schapery and S.L. Sicking

Thermoelastic Martensitic Transformation Induced Plasticity -
    Micromechanical Modelling, Experiments and Simulations   77
       HWANG Keh-Chih and SUN Qing-Ping

Damage Analysis of Brittle Disordered Materials: Concrete and Rock   101
       J.G.M. van Mier, E. Schlangen, A. Vervuurt and M.R.A. van Vliet

Relations between the Mechanical Behaviour of Polymers and their
    Processing Methods and Conditions   127
       W. Michaeli

Structural Integrity Assessment of High Integrity Structures and
    Components: User Experience   143
       P. E. J. Flewitt

Foreword

The International Conference on Mechanical Behaviour of Materials, ICM, has been organised to take place in The Hague, The Netherlands, from May 28 till June 2, 1995. Previous editions were successfully organised in Kyoto (1971), Boston (1975), Cambridge (1980), Stockholm (1983), Beijing (1987) and again Kyoto (1991). This Seventh Conference, ICM-7, has been organised by the European Structural Integrity Society, ESIS, in co-operation with Delft University of Technology.

The Conference program reflected the progress made in science and technology of mechanical behaviour of materials since the last Conference in 1991. From the abstracts that were submitted it appeared that a large part was focused on the micro-mechanical and constitutive modelling of various types of mechanical behaviour: deformation (plasticity, creep), damage (fatigue, creep, creep-fatigue interactions) and fracture (brittle and ductile). It was therefore decided to focus on modelling in the plenary sessions with invited contributions. This book presents the full length articles of seven of the plenary lectures of ICM-7.

Five of the seven articles deal with the micro-mechanical and constitutive modelling of the mechanical behaviour of various material classes, including:

- (Poly-)crystalline materials;
- Composites with viscoelastic matrix (e.g. polymer matrix composites);
- Brittle disordered materials (e.g. concrete, rock, non-transformable ceramics);
- Toughened ceramics and shape memory alloys.

The remaining two papers, although in a broad sense also dealing with modelling, focus on two specific aspects.
The first deals with the influence of processing methods and conditions on the mechanical properties of plastic products, a very important subject for the fabrication of plastic mass-products. The second deals with structural integrity assessment of structures and components, a subject that should not be lacking in a conference organised by the European Structural Integrity Society, but also an important aspect for designers and operators of high-integrity structures and components.

To have this book available at the conference required quite some discipline from the authors. I wish to thank them for submitting their manuscripts in time to allow for a reviewing process in which each paper was reviewed by at least two, but most times three reviewers. Also the revised manuscripts were received in time, and most times in a format that minimised the effort to bring all articles in a uniform format.

Finally I wish to thank the reviewers of the papers. The list is to long (20 persons) to address them personally, but their work was a major contribution to the realization of this book.

A. Bakker
Editor

April 1995

*A. Pineau**

Effect of Inhomogeneities in the Modelling of Mechanical Behaviour and Damage of Metallic Materials

Reference: Pineau, A. (1995), Effect of Inhomogeneities in the Modelling of Mechanical Behaviour and Damage of Metallic Materials. In: *Mechanical Behaviour of Materials* (ed. A. Bakker), Delft University Press, Delft, The Netherlands, pp. 1-22.

Abstract: Metals and alloys contain many sources of inhomogeneity. They may either pre-exist in the material or develop under deformation. This non-uniformity in the local properties of the materials may strongly affect their overall stress-strain response and, in particular, their damage behaviour. In this paper an overview is given of a number of theories which have been introduced in the literature and which provide guidelines for the modelling of the mechanical behaviour of these heterogeneous materials.

The paper is divided into two parts. The first part, which deals with the mechanical behaviour of non-damaged materials, introduces a general scheme adopted to predict the overall behaviour of a polycrystal from knowledge of the behaviour of individual single crystals. In this scheme we concentrate on three aspects: intergranular and transgranular hardening and anisotropy effects due to crystallographic texture. In each case an attempt is made to show the scale at which the effect of non-uniformity occurs. In the second part of the paper, which is devoted to damaging materials, two types of models are introduced : (i) those in which the constitutive equation of the material is not coupled with the damage evolution and (ii) those in which a strong coupling is introduced through appropriate constitutive equations such as the Gurson-Tvergaard-Needleman potential. Both types of models are applied to a specific material, a cast duplex ($\alpha + \gamma$) stainless steel. In this material damage initiates from cleavage cracks formed in the embrittled ferrite (α) phase. These microcracks lead to the formation of cavities which grow in the austenite (γ) phase and eventually coalesce to final fracture. Monte Carlo type numerical simulations are introduced to reproduce the strongly non-uniform distribution of the nucleation rate of cleavage microcracks observed in this material. It is thus shown that a fully coupled model accounting for the damage non-uniformity is able to simulate all stages of ductile damage, including crack initiation and crack growth.

1. Introduction

The modelling of the mechanical behaviour of materials, together with the evolution of the damage when they are deformed, is a research field which remains extremely active. Several approaches must be distinguished : (i) the microscopic theories, (ii) the macroscopic theories and (iii) the "micro-macro" approaches which are now being largely developed.

The direct relationship between the details of microstructure and the mechanical state of crystalline solids has been and remains the subject of a great deal of discussion. "Microscopic" theories are often theories for strain-hardening, i.e. the increase in flow stress with plastic strain at a prescribed strain rate. These theories have proceeded on the basis of models for the microstructure with relatively little input from the mechanical data. In these theories the major experimental guide for quantitative analysis is electron microscopy, which measures a scalar ρ, the dislocation density which prescribes an isotropic kind of hardening and is unsuited for describing anisotropic effects (see, for example, Asaro, 1975). Moreover, in most cases, only monotonic uniaxial mechanical tests are performed to test these microscopic theories, so that it is impossible to differentiate the isotropic and kinematic components of the strain hardening.

---

* *Centre des Matériaux, École des Mines, B.P. 87 - 91003 EVRY Cédex, France*

Continuum plasticity theories, on the other hand, have generally ignored microstructural details, but they are more complete in their descriptions of the mechanical behaviour. In particular, they are able to describe much more complex multi-axial loadings. However, these phenomenological macroscopic theories use a large number of material coefficients to represent complex mechanical behaviour. The identification of these coefficients requires numerous often quite sophisticated tests. Moreover, these macroscopic theories give no insight into the physical processes controlling the deformation and the accumulation of damage.

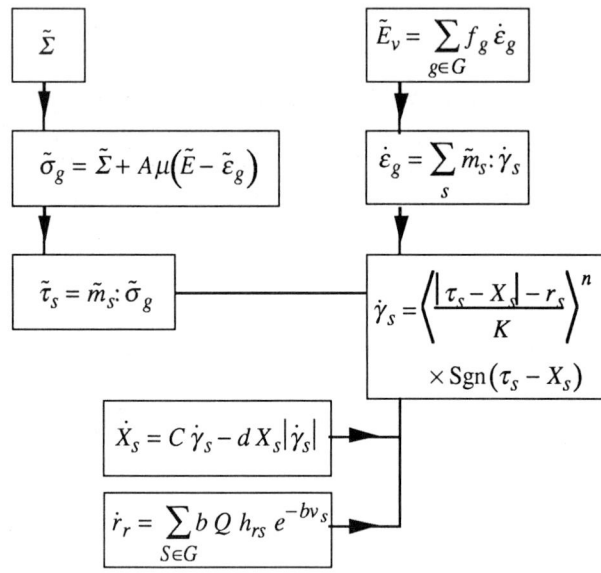

Fig. 1 General scheme for modelling the plastic behaviour of polycrystals from the behaviour of single crystals.

There is a third class of models, which are sometimes referred to as "micro-macro models". In these models, which are now being widely developed (see, for example, Mécamat, 1993), plasticity is considered as the result of various micromechanisms which take place on discrete active slip systems. Homogenization techniques and, in particular, the self-consistent formulation have provided the general frame for establishing these models. A general description of such models, proposed by Cailletaud (1988, 1992), is given in Figure 1. The bases of these micro-macro models are briefly described in the introduction to the present paper, after which we shall concentrate on two important aspects, the intergranular hardening and the transgranular hardening respectively, in relation to the effect of microstructural inhomogeneities on the mechanical behaviour of metallic materials. The aim of this first part of the paper is to illustrate by means of a number of selected examples how the inhomogeneity of plastic deformation can influence the mechanical response of crystalline materials and to show how such a micro-macro model can provide guidelines for the modelling of a number of types of mechanical behaviour. In the second part of the paper the aim is to show that the effects of non-uniform distribution of damage are even more important in order to describe and model the mechanical response of the materials. In this part we will strongly rely upon a recent study carried out in the author's laboratory and dealing with duplex (austenite + ferrite) stainless steels.

Two critical steps must be examined carefully in Figure 1 : (i) the localization process, i.e. the method of calculating local stresses $\tilde{\sigma}_g$ and local strains $\tilde{\varepsilon}_g$ from the macroscopic stress $\tilde{\Sigma}$ and the macroscopic strain $\tilde{E}$, which is, in particular, the basis for intergranular hardening effects, and (ii) the local description of transgranular hardening on each slip system in a given grain.

In the localization process, according to the self-consistent theory developed in particular by Kröner (1958), the local stress is related to the macroscopic stress by the following expression :

$$\tilde{\sigma}_g = \tilde{\Sigma} + A\mu(\tilde{E} - \tilde{\varepsilon}_g) \qquad (1)$$

where A is termed the plastic constraint factor of the matrix and μ is the shear modulus. A takes different values, according to various theories : $A = \infty$, in the Taylor model (1938); $A = 2$ in the Lin model (1957); $A = 2(1-\beta)$ in the Kröner model. In the latter model β is a function of the grain shape. For a spherical inclusion $\beta = 2(4-5\nu)/15(1-\nu)$. A more rigorous relation was formulated by Hill (1965), who applied the Eshelby solution (1957) in the context of inhomogeneity problems. Under monotonic, proportional loading the Hill solution has been modified by Berveiller and Zaoui (1979) to give a simpler form:

$$\tilde{\sigma}_g = \tilde{\Sigma} + 2\mu\alpha(1-\beta)(\tilde{E} - \tilde{\varepsilon}_g) \qquad (2)$$

where α is called the plastic accommodation function. The value of α depends on the actual secant shear modulus, μ, and the secant Poisson ratio. Typically after a few per cent of plastic strain α is of the order of 10^{-2}.

Once the local stress is calculated, $\tilde{\sigma}_g$ is then resolved into the shear stress on each slip system. In the Cailletaud model, which is formulated as a viscoplastic model, the shear strain rate $\dot{\gamma}_s$ on each slip system is related to the "effective stress" by a power law function, as indicated in Figure 1, where X_S and r_S represent the kinematic and the isotropic hardening respectively. Evolution laws, similar to those adopted by a number of authors who have developed phenomenological models (see e.g. Lemaitre and Chaboche (1985)), are used for the kinematic and the isotropic hardening. In the isotropic hardening an interaction matrix between the different slip systems is also introduced. Once the local shear strain rate is known, the local strain rate $\dot{\varepsilon}_s$ is calculated and then averaged to determine the macroscopic strain rate, as indicated in Figure 1.

2. Mechanical Behaviour

In this part we concentrate on three aspects: intergranular and transgranular hardening and anisotropic effects due to crystallographie texture, which are included in the general scheme shown in Figure 1. Firstly details are given of three examples illustrating the importance of the intergranular hardening effect. In selecting these examples dealing with fracture an attempt is made to show the importance of the inhomogeneity of plastic deformation in polycrystals and multiphase materials. The transgranular hardening effect is examined with the emphasis laid on the micromechanisms which develop a strong kinematic hardening effect. The part of the paper devoted to the mechanical behaviour of materials is concluded by presenting experimental results on the effect of crystallographic texture on the plastic anisotropy of Al alloys and the modelling of this effect with the micro-macro approach shown in Figure 1.

2.1. Intergranular hardening

Equation (1) simply indicates that a component additional to the applied stress is active locally when the local strain, $\tilde{\varepsilon}_g$, is different from the mean strain or the overall strain, $\tilde{E}$. This condition is met in ductile fracture initiated from inclusions, the first example selected, also in the deformation of polycrystals, which under certain circumstances is accompanied by brittle intergranular fracture, the second example used here to illustrate the effect of non-uniform distribution of deformation, and again in the deformation of dual phase materials, the third example.

2.1.1. Initiation of cavities from inclusions

It is well established that inclusions play an important role in ductile rupture. The discontinuous nucleation of cavities, which occurs at large (≈ 1 μm) and widely spaced inclusions, can be described in terms of continuum mechanics. Argon and his co-workers (1975) proposed a nucleation model based on the existence of a critical stress, σ_d, such that:

$$\sigma_d = \sigma_{eq} + \sigma_m \qquad (3)$$

where σ_{eq} is the local equivalent Von Mises stress and σ_m is the hydrostatic stress. The inhomogeneity in plastic deformation between the matrix and the particles does not appear explicitly in Equation (3). This is the reason why the feature was more clearly defined in the theory developed by Beremin (1981) to account for cavity nucleation at MnS particles in low-alloy steels. An expression was used, directly derived from an application of the inclusion theory by Eshelby (1957), which is one of the bases of the self-consistent formulation, i.e. :

$$\sigma_d = \Sigma_1 + k(\sigma_{eq} - \sigma_o) \qquad (4)$$

where Σ_1 is the maximum principal stress, σ_o is the yield strength and k is a function of particle shape. The data given in Figure 2 show that Equation (4), which is very similar to Equation (1), accounts very well for the test results obtained at different temperatures. It is also noticed that the values of k and those of σ_d are different when the material is tested along two different directions. The different values of k are related to the anisotropy in the shape of the MnS particles, elongated in the rolling direction, whereas the different values for σ_d correspond to different mechanisms of fracture. In the longitudinal direction, σ_d corresponds to the fracture of MnS inclusions, whereas in the transverse direction, σ_d is related to the decohesion of the inclusions from the matrix.

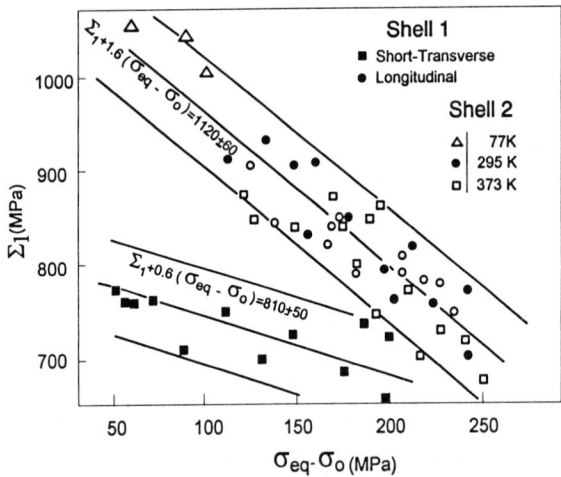

Fig. 2 Cavity nucleation from MnS inclusions in two heats of A508 steel. Effect of temperature and loading orientation, (Beremin, 1981).

2.1.2. Brittle intergranular fracture in a temper-embrittled C-Mn-Ni-Mo steel

The classic example of intergranular brittle fracture occurs in temper-embrittled low-alloy steels caused by the segregation at grain boundaries of impurities such as S, P, As, Sb, etc... The ductile-brittle transition temperature is shifted to higher temperatures as a consequence of the intergranular embrittlement. The magnitude of the effect is dependent on the quantity and type of segregated solute atoms. The mechanical modelling of the embrittlement and the associated loss in intergranular cohesion has received relatively little attention. There has been recent renewed interest in this phenomenon (see Cottrell, 1989, 1990a; 1990b).

Temper embrittlement has recently been studied by testing smooth and notched specimens of a C-Mn-Ni-Mo steel at different temperatures (Kantidis et al, 1994a and b). The results obtained on notched specimens as shown in Figure 3a indicate that the intergranular fracture stress tends to increase with temperature. Moreover, a wide scatter in the results is noted. A model was proposed to account for this temperature dependence and scatter (Kantidis et al, 1994a), but one may wonder whether the temperature is the appropriate variable. Considering the mean strain at fracture, it is observed that the ductility is an increasing function of temperature, as expected (Figure 3b). In a polycrystal the inhomogeneity in plastic strain between the different grains is also expected to decrease with increasing overall plastic strain. This was shown recently using the model presented previously, which was applied to simulate the deformation of polycrystalline specimens of a 2024 Al alloy (Pilvin, 1993). The results reported in Figure 4 clearly show that the histograms giving the relative distribution of plastic strain within the grains become increasingly sharper when the plastic strain is increased. This means that qualitatively, according to Equation (1), the contribution of the intergranular hardening effect is a decreasing function of applied plastic strain. In other words, there is a large probability that at low applied strain the local stress responsible for brittle intergranular fracture is larger than the applied stress. A quantitative model has to be developed to take into account this inhomogeneity in the distribution of plastic strain between the different grains of a polycrystal, in order to predict the temperature dependence of the intergranular fracture stress.

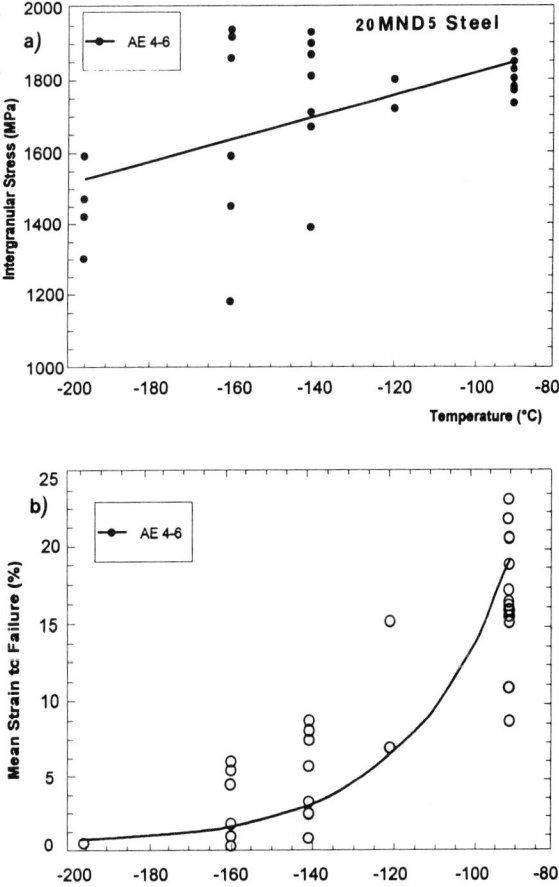

Fig. 3 Intergranular fracture in 20MND5 steel. Intergranular fracture stress (a) and strain to failure (b) versus temperature (Kantidis et al., 1994a).

2.1.3. Nucleation of cleavage cracks in the ferrite of duplex stainless steels

Cast duplex stainless steels are two-phase (ferrite α + austenite γ) materials which may contain up to 30% of ferrite. Aging these materials around 300°C-400°C produces a considerable decrease in their fracture toughness due to the formation of cleavage cracks in the ferrite phase after deformation. Aging also produces a significant increase in the hardness of the ferrite phase due to

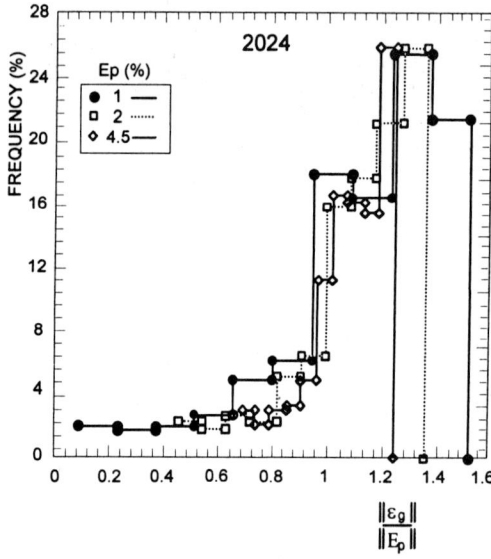

Fig. 4 Histograms showing the calculated distribution of plastic strain within the grains of a polycrystal of 2024Al alloy, ε_g, as a function of the macroscopic applied strain, E_p.

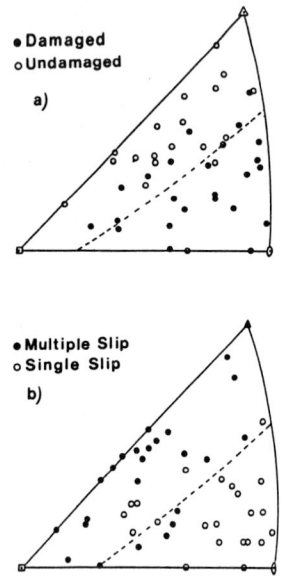

Fig. 6 Cast duplex stainless steel. a) Position of the tensile axis in the γ standard triangle; b) Observation of single or multiple slip in the γ standard triangle.

the ($\alpha + \alpha'$) decomposition of these Cr-rich ferrites (Joly, 1992). The damage mechanisms of these materials are described in more detail in the second part of this paper.

One heat of duplex stainless steel was investigated containing 25% ferrite, aged at 400°C for 700 hours. Interrupted tests carried out on tensile specimens showed that the damage corresponding to the formation of cleavage cracks was not homogeneously distributed. An example illustrating this pronounced inhomogeneity in the distribution of cleavage cracks is shown in Figure 5, where the damage is analysed using Voronoï cells. Figure 5b and 5c clearly show that the cleavage cracks are grouped into clusters. The importance of this inhomogeneity for modelling the fracture of these materials will be stressed in the second part of the paper. Here we would like to present only a number of results showing that the clustering of cleavage cracks is directly related to an "intergranular" type hardening effect.

Detailed metallographical analysis showed that this clustering effect was not related to a specific orientation of the ferrite phase but rather to the orientation of the soft phase, i.e. the austenite (Joly et al, 1990). This is shown in Figure 6, where it is observed that cleavage is almost completely suppressed when the tensile axis is located in a band close to $(100)_\gamma$ - $(111)_\gamma$ zone. Figure 6b shows that this zone also corresponds to orientations for which multiple slip is observed. On the other hand, cleavage of the ferrite phase is favoured when the austenite surrounding these islands of ferrite is oriented to be in single slip. A simple explanation for the observed effect is the following. Single slip is accompanied by a rotation of the austenite grains, but since these grains are constrained by the ferrite network, this rotation cannot easily take place. This produces an elevation of the local stresses

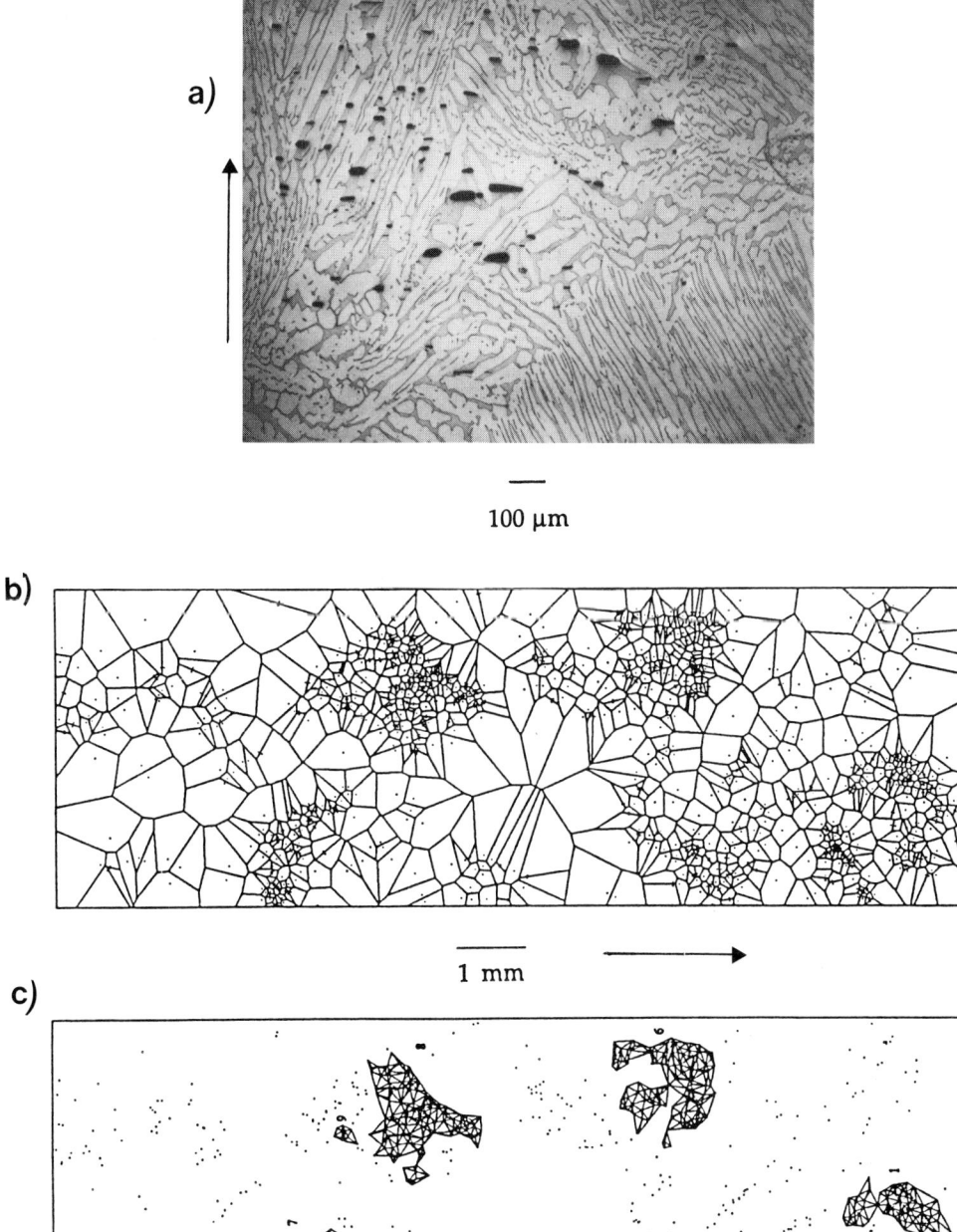

Fig. 5 Cast duplex stainless steel. a) Optical micrograph illustrating the nonuniformity in the distribution of cleavage cracks in the ferrite phase; b) Voronoi cells showing the formation of clusters of cleavage cracks which are isolated in c). The arrow indicates the direction of the tensile axis.

in the ferrite, which are partly relaxed by the formation of cleavage cracks. This example illustrates the importance of intergranular stresses, which are due this time not only to the pronounced inhomogeneity in the distribution of plastic deformation between the austenite and the ferrite, but also to the slip multiplicity in the soft phase.

2.2. Transgranular hardening

The importance of transgranular hardening, in particular the kinematic hardening, is illustrated by two examples dealing with the cyclic behaviour of : (i) precipitation-strengthened materials and (ii) fully pearlitic steels.

2.2.1. Shape of hysteresis stress-strain loops in precipitation strengthened materials in relation to kinematic hardening

Figure 7 shows the shape of these hysteresis loops in Waspaloy a nickel base superalloy strengthened by the precipitation of the γ' phase. Cyclic tests were performed at the same total strain on specimens which were given different heat-treatments, i.e. underaging and overaging corresponding to particle sizes of 8 and 90 nm respectively. Small γ' particles ($\phi = 8$ nm) are sheared by dislocations, whereas the large ones ($\phi = 90$ nm) are by-passed by the Orowan process. The heat treatments were selected in such a way that these materials exhibit the same monotonic yield strength and almost the same cyclic stress as observed in Figure 7. This figure clearly shows that the hysteresis loop is more square-shaped in the underaged than in the overaged condition. The Bauschinger effect, i.e. the kinematic hardening effect, is thus more pronounced when the particles are bypassed by dislocations.

Several forms of kinematic hardening related to dislocation micromechanisms were introduced by Asaro (1975). The case illustrated in Figure 7 corresponds to the third type, i.e. the micromechanisms are perfectly reversible. A sketch indicating the sequence of interactions between the particles and the dislocations is shown in Figure 8. This sketch has received some experimental confirmation with in-situ T.E.M. observations (Louchet, 1992). The interaction is perfectly reversible since, when the dislocation moves back, it reacts with the loops left around the particles and anneals them. During its backward motion the dislocation is submitted to an attraction force from the loops left around the particles, which leads to easy plastic deformation each time the stress is reversed.

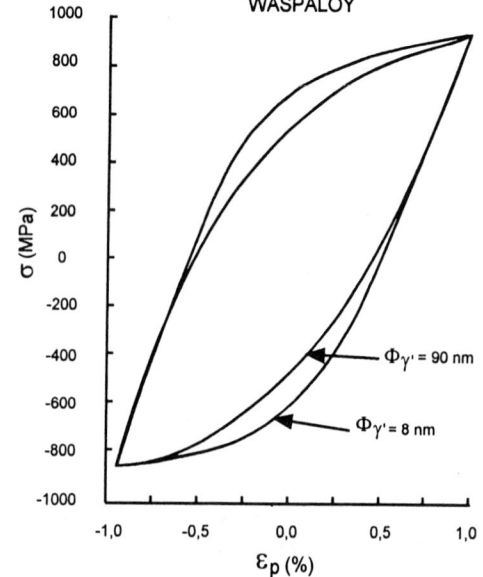

Fig. 7 Cyclic stress-strain curve in Waspaloy with two sizes of γ' particles.

2.2.2. Cyclic stress-strain behaviour of fully pearlitic steels

In dual-phase materials subject to large cumulative strains the kinematic hardening effect is expected to be strongly affected by the microstructural details when the "relaxed" dislocation substructures, for instance the dislocation cells formed in the matrix in the absence of the second phase, are larger than the mean free path between the particles. This situation was observed in a recent study devoted to the cyclic behaviour of eutectoid steels (Jeunehomme, 1991). Two fully pearlitic steels were submitted to various heat treatments to produce large variations in cementite interlamellar spacing. Low-cycle fatigue tests were performed at room temperature on these materials. The amplitude of the Bauschinger effect was measured by the ratio $X/(\Delta\sigma/2)$, as indicated in the insert of Figure 9, where the cyclic stress-strain curves corresponding to two values of the interlamellar spacing, Sp, are reported. In the material with the large value of Sp (0.27 μm) the cyclic curve is continuous, with a unique value of the cyclic work hardening exponent in the range of plastic strain which was investigated. In the small interlamellar spacing material ($Sp = 0.10$μm) a different situation is observed. At large plastic strain amplitudes ($\approx 0.1\%$) the slope of the cyclic σ–ε curve is the same as the slope measured with large Sp material. In both cases the ratio $X/(\Delta\sigma/2)$ is close to 0.50 - 0.60. Figure 9 shows that at lower plastic strain the cyclic stress is anomalously high. This modification in the behaviour of the material is accompanied by an increase in the $X/(\Delta\sigma/2)$ ratio, which becomes larger than 0.70, denoting a pronounced Bauschinger effect.

T.E.M. observations showed that the dislocation microstructures were different from each side of the transition on the cyclic σ–ε curve. At large plastic strain, dislocation cells were observed, whereas at lower plastic strain amplitude these cells did not

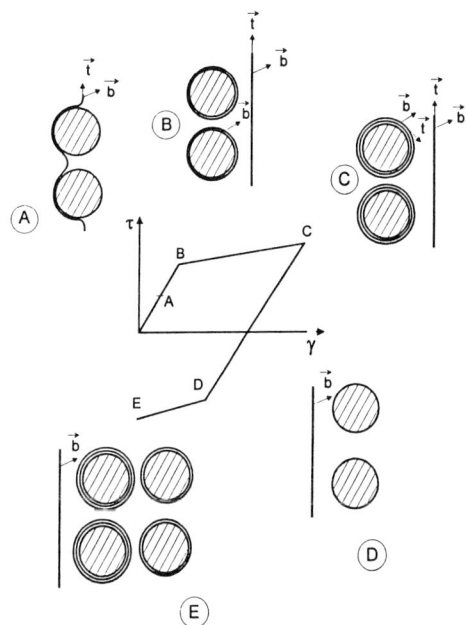

Fig. 8 Sketch showing the successive positions of a dislocation along a stress-strain curve in a material in which the particles are by-passed.

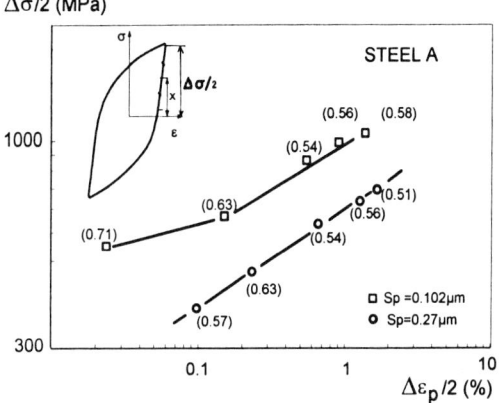

Fig. 9 Cyclic stress-strain curves in two pearlitic steels with different values for the interlamellar spacing, Sp. The numbers indicate the value of the Bauschinger effect measured by the ratio $X/(\Delta\sigma/2)$.

appear but the dislocations were confined within the walls formed by the cementite platelets. Similar behaviour was shown by Sunwoo et al (1982), as indicated in Figure 10. One can say that in these materials, for a given interlamellar spacing, there is a critical plastic strain amplitude below which the inhomogeneity in plastic deformation is confined within the walls formed by the cementite platelets. Under these conditions most of the deformation is accommodated by the geometrically necessary dislocations introduced by Ashby (1971). This produces a large kinematic hardening effect, as expected. It may also be added that the fatigue life was also largely influenced by these microstructural features. In particular, the slope of the Manson-Coffin plot was shown to be larger than the slope usually observed in most materials, which is -0.50 (see Figure 11). A close examination of the endurance curves also showed the existence of a transition which corresponds to the transitional behaviour observed in the dislocation microstructure. An example of such a situation is given in Figure 11b for the fine pearlite material. The comparison with Figure 9 shows that this transition in the slope of the Manson-Coffin law coincides with that observed for the cyclic σ–ε curves. This example shows that microstructural modifications may affect not only the constitutive equation of the materials but also their damage behaviour.

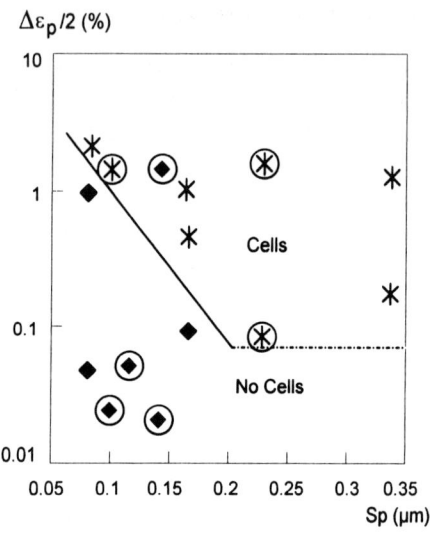

Fig. 10 Fully pearlitic steels. Results showing that the strain amplitude, $\Delta\varepsilon_p/2$ in a cyclic test necessary for dislocation cell formation decreases when the interlamellar spacing increases. (Sunwoo et al., 1982). The circles correspond to the results obtained by Jeunehomme (1991).

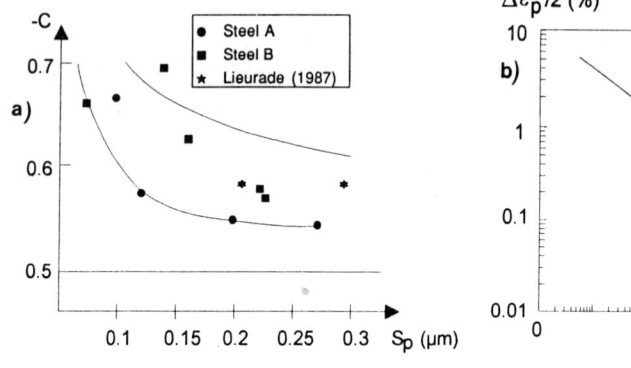
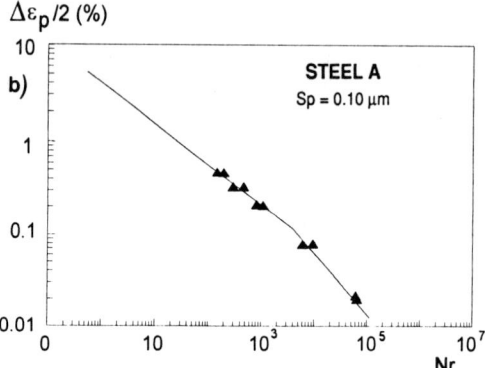

Fig. 11 Fully pearlitic steels. a) Variation of the c exponent in the Manson-Coffin law $\Delta\varepsilon_p/2 = \varepsilon'_f(N_r)^{-c}$ where ε'_f is a constant and N_r the number of cycles to failure, with the interlamellar spacing, Sp; b) Variation of the fatigue life N_r with strain amplitude, $\Delta\varepsilon_p/2$ in one specific heat (Steel A).

2.3. Plastic anisotropy of aluminium alloys

The model described in the introduction was recently used to predict the plastic anisotropy of 7xxxAl alloys (Achon, 1994). The results are presented in order to illustrate the effect of another type of inhomogeneity related to the anisotropy. Two materials taken from thick plates (≈ 30 mm), a 7075 and a 7475 alloy, were investigated in three conditions (T3, T6 and T7). Tensile specimens were machined in six directions. Besides the three main directions of the thick plates, i.e. the longitudinal (L), the transverse (T) and the short transverse (S) directions, three other directions were investigated at 45 degrees to the main directions. These directions were denoted SL, ST and LT, with LT signifying that the tensile axis was in the LT plane at 45 degrees to the L and T directions. The Lankford coefficient, defined as the ratio between the plastic strain measured in the transverse direction of the tensile specimen and the longitudinal strain, was measured and compared with the results obtained from the application of the model (Figure 12).

The material was modelled as a polycrystal of more than 400 grains oriented to reproduce the measured crystallographic texture, as indicated in Figure 12a, which corresponds to a (111) pole figure. This orientation of the grains is taken into account in the model through the orientation tensor $\tilde{m}_s$ introduced in Figure 1 and the averaging procedure used to calculate the macroscopic strain, $\tilde{E}$, from the microscopic shear strain or shear strain rate, $\dot{\gamma}_s$. The results obtained from this "microscopic" polycrystalline model are labelled {p}. The plastic anisotropy of these Al alloys was also described in terms of conventional anisotropic plasticity by the Hill model (1950). The results obtained with this "macroscopic" model are indicated by {H}. Figure 12 shows that the polycrystalline model is able to reproduce the anisotropy in plastic deformation of these FCC materials by using as an input the initial crystallographic texture, which is a measure of the deviation from a homogeneous isotropic condition.

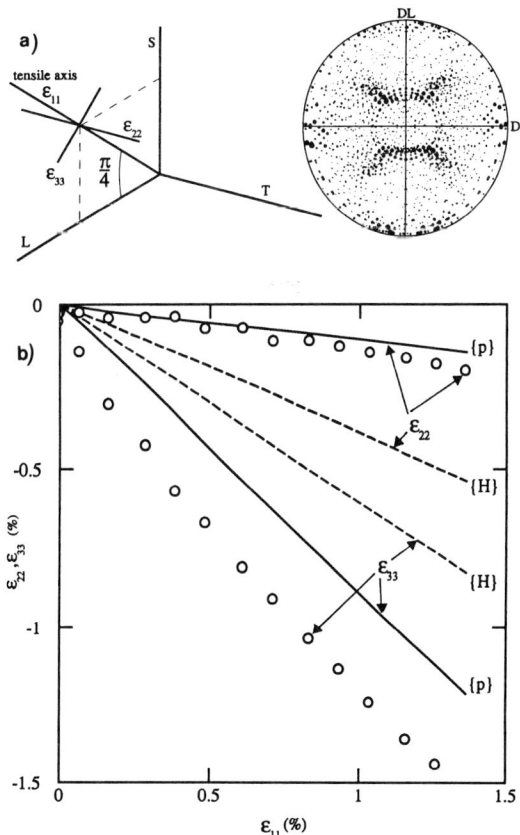

Fig. 12 Anisotropy in plastic deformation in a 7075 Al alloy. a) Position of the tensile axis and of the transverse strains, ε_{22}, ε_{33} and (111) pole figure; b) Variations of ε_{22} and ε_{33} as function of the axial tensile strain, ε_{11}. Comparison between the experiments (circles) and two models (see text).

2.4. Partial conclusions

This broad overview of the possibilities offered by the so-called "micro-macro" approaches to model the mechanical behaviour of metallic materials has shown that it is now possible to model complex mechanical properties of polycrystals, including the effect of local stresses on fracture, the Bauschinger effect in relation to the interaction of dislocations with a number of microstructural details and the effect of crystallographic texture.

However, this optimism must be tempered for one main reason, springing from the fact that in this type of homogenization technique the geometrical characters of the material or the plastic deformation are not taken into account. In particular the grain size dependence of the mechanical properties of the materials cannot be explained by this type of model. Moreover, the inhomogeneity of plastic deformation within a grain, i.e. the formation of discrete slip bands, which is known to play a dominant role in a number of problems, is not considered. The absence of any geometrical factor in this type of theory leads thus to a number of serious limitations.

3. Damage

3.1. Introduction

In this part of the paper an attempt is made to indicate how it is possible to model the damage of metallic materials. As indicated previously, this part is essentially devoted to the description and modelling of damage in cast duplex stainless steels. Two main types of models, representative of those usually used in other materials, are introduced : (i) those in which there is no coupling or only partial coupling between the constitutive equation of the material and the damage evolution, and (ii) those in which this coupling effect is taken into account. It should also be added that only ductile damage associated with the nucleation, growth and coalescence of cavities is considered here.

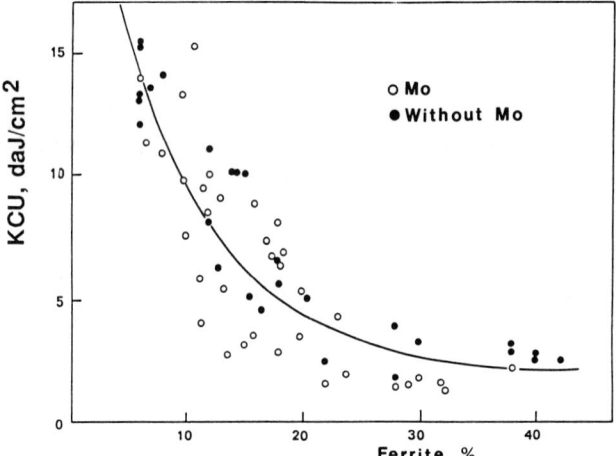

Fig. 13 Variation of the Charpy fracture toughness measured at 20 °C in cast duplex stainless steels aged at 400 °C for 10.000 hours as a function of the ferrite content (Bonnet et al., 1990).

Cast duplex stainless steels are two-phase materials containing up to 30% of BCC ferrite in an FCC austenitic matrix. These materials are used for the manufacture of a number of components, in particular the pipes and elbows used in electronuclear reactors. They may then be submitted to an embrittlement effect due to prolonged exposure at in-service temperatures close to 300°C. It has been shown that after accelerated aging at 400°C the ferrite phase is hardened and embrittled by a phenomenon similar to the "475°C embrittlement", which is well documented in Cr-rich ferritic steels. This produces a considerable decrease of the Charpy toughness, as shown in Figure 13 (Bonnet et al, 1990; Meyzaud et al, 1988). Charpy impact properties and fracture toughness of aged material present an important scatter, which is reduced when large specimens are used (Meyzaud et al, 1988). The modelling of damage in these materials must therefore be able to describe not only the variation in the ductility with aging conditions but also the scatter and the size effect.

We first present the results of a number of metallographical observations which were made in order to investigate the micromechanisms of damage in these materials. These results were obtained from one specific heat, taken from a centrifugally cast pipe which was aged at 400°C for 700h. Only a brief account of these results is given here. Full details can be found elsewhere (Joly et al, 1992, Joly et al, 1993). Next the mechanics of porous materials is briefly introduced. This is the basis of the coupling between the constitutive equations and the damage. Finally both types of models, with and without coupling, are applied, using the results of metallographical observations to predict the ductility and fracture toughness of these materials.

3.2. Metallographical observations

As already illustrated in Figure 5a, cleavage cracks are initiated in the ferrite phase and preferentially in areas such that the surrounding austenite is deformed in single slip (Joly, 1992). The number of grains containing the clusters of cracks was measured (≈ 5.7 grains/cm^2). Their size was found to be of the order of 2 mm along the longitudinal axis of the tube and 0.75 x 1 mm^2 in a planar section perpendicular to this axis. It was shown that the onset of crack nucleation was not dependent on the applied maximum principal stress but rather on the achievement of a critical plastic strain of the order of $2 \ 10^{-2}$ (Joly et al, 1992). Measurements of the number of cracks per unit area showed that the density of cracks increased in the preferentially oriented grains and that new damaged grains appeared with an increase in plastic strain. Moreover, measurements of the area of cleavage cracks at different levels of plastic strain indicated that the distributions of the surface area of these cracks were not modified. This indicates that once cleavage cracks are initiated they immediately reach their maximum extension.

These cleavage microcracks initiate the formation of cavities which grow by plastic blunting in the austenite, as shown previously (Pineau, Joly, 1991). Macroscopic fracture initiation occurs by cavity coalescence, which takes place within one cluster of microcracks.

In the next section it is assumed that a flat crack produces the same softening effect as a spherical cavity with a radius $R = \sqrt{A/\pi}$, where A denotes the surface area of the crack. This assumption, which has recently received some theoretical support (Gologenu et al, 1994), was made in order to relate the nucleation rate of cleavage cracks with plastic strain to the increment of the volume fraction of porosity, df_n, which was expressed as :

$$df_n = A_n d\varepsilon_{eq} \qquad (5)$$

where A_n is the nucleation rate and $d\varepsilon_{eq}$ is the increment of the Von Mises strain. The nucleation rate, A_n, was assumed to be distributed from grain to grain, as indicated previously. In particular A_n is higher within grains which are suitably oriented. In those grains, A_n was found to be close to one, while the mean value of A_n was close to 0.50. The distribution of A_n was carefully determined from detailed metallographical observations (Joly, 1992).

3.3. Mechanics of porous materials

As shown in the previous section, although the aged material is very brittle, the micromechanisms of fracture are similar to those associated with ductile fracture, by nucleation, growth and coalescence of cavities. Ductile fracture is usually considered as an instability due to the softening effect induced by cavity growth (Berg, 1969). Several potentials have been introduced to model the plastic behaviour of porous materials (see e.g. Gurson, 1977 and Rousselier, 1981 and 1987). In particular in the Gurson potential modified by Tvergaard (1981) the effect of a volume fraction of cavity, f, can be accounted for by a yield criterion expressed as :

$$\frac{\sigma_{eq}^2}{\sigma_Y} + 2q_1 f \cosh\left(q_2 \frac{3}{2} \frac{\sigma_m}{\sigma_Y}\right) - 1 - q_3 f^2 = 0 \tag{6}$$

where σ_{eq} is the Von Mises equivalent stress, σ_m is the mean stress and σ_Y is the flow stress of the matrix. The coefficients have the following values $q_2 = 1$, $q_3 = q_1^2$ and $q_1 = 1.50$. It was shown by Perrin and Leblond (1990) that $q_1 = 4/e \approx 1.47$. It can easily be shown that, neglecting the second order term $q_3 f^2$, the dilatation calculated from the above expression can be expressed in terms of porosity increment as :

$$df_G = \frac{3}{2} f q_1 \sinh\left(\frac{3}{2} \frac{\sigma_m}{\sigma_Y}\right) d\varepsilon_{eq} \tag{7}$$

Berg (1969) proposed a ductile fracture criterion in terms of plastic instability, while Yamamoto (1978) used a similar approach for materials following the Gurson potential. According to these theories, the softening effect and dilatation effect associated with cavity growth favours the localization of deformation along planar bands. A simplified criterion for strain localization was introduced by Mudry (1982). It is assumed by this author that fracture occurs in a volume element when the work-hardening rate of the matrix is exactly balanced by the softening effect associated with the cavities, i.e. when :

$$\frac{d\sigma_{eq}}{d\varepsilon_{eq}} = 0 \tag{8}$$

The influence of the nucleation rate of cavities, A_n, in Equation (5), on the stress-strain curves calculated from Equation (6) for a stress triaxiality σ_m/σ_Y equal to 0.60 is shown in Figure 14. A strong influence of the nucleation rate is noticed. Similarly the variation in the strain to failure, ε_f, predicted from Equation (8) is shown in Figure 15 as a function of stress triaxiality and A_n. In this figure it is noticed that ε_f becomes less dependent on stress triaxiality ratio as A_n increases, as expected, since the nucleation law given by Equation (5) is not dependent on stress triaxiality, which is opposite to the situation encountered when fracture is essentially controlled by the growth of cavities. In cast duplex stainless steels it was shown (Joly, 1992) that the ductility was essentially controlled by the continuous nucleation of cleavage cracks.

3.4. Modelling of ductility

The model which has been presented above applies to a situation where the nucleation rate, A_n, is spatially uniformly distributed. This is not the situation in the present material. This non-uniform distribution of the clusters of microcracks is responsible for the scatter in the results of mechanical tests and for the size effect as shown below. Two types of model can be used to account for this inhomogeneity in the distribution of the damage. In the first type, the calculation of the stress-strain distribution in a notched specimen or component is made without accounting for the coupling effect between the constitutive equation and the damage. The clusters are distributed within the structure and a Monte Carlo type simulation is used as a post-processor routine to predict the failure conditions. In the following this type of model is referred to as "uncoupled". In the second type of model, the structural analysis is fully coupled with the calculation of the evolution of damage.

3.4.1. Uncoupled modelling

A statistical analysis of the failure of axisymmetric notched specimens was carried out by finite element calculations. Notched tensile specimens with a minimum diameter of 10 mm and notch radii of 10, 4 and 2 mm were used. These specimens were calculated with the stress-strain curve determined on conventional tensile specimens. In these specimens the damage is so "diluted" that it can be considered that this stress-strain curve represents within a first approximation the mechanical response of the undamaged material.

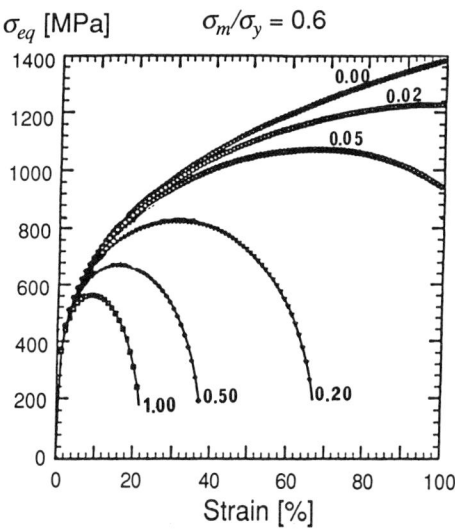
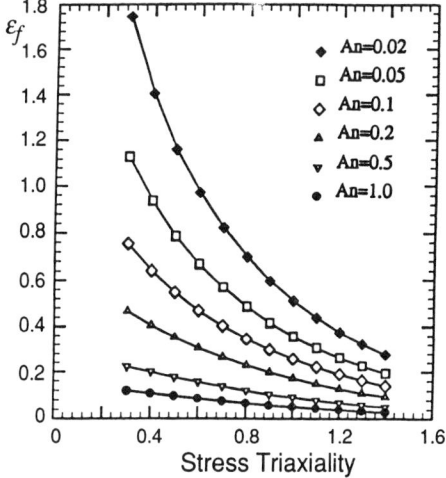

Fig. 14 Numerical simulations showing the influence of the nucleation, rate A_n, on the stress-strain curve of a Gurson-Tvergaard damaging material. A_n-values as indicated per curve.

Fig. 15 Variation of the ductility with stress triaxiality ratio for different values of the nucleation rate, A_n.

A large number of parallelepipedic grains, with dimensions similar to those determined experimentally (2 x 1 x 0.75 mm²) were randomly generated in a large volume. Their coordinates were obtained by a random number generating routine. For each grain a nucleation rate, A_n, was generated by a Monte Carlo method, according to the distribution function determined experimentally. Samples of 100 specimens of each notched geometry were located in this volume, each of them including a few grains. Failure analysis was carried out by a post-processor routine at every Gauss point and at every load increment. If the Gauss point is located within a cluster, its nucleation rate, A_n, is used to integrate Equation (5) and Equation (7). Full details are given elsewhere (Joly et al, 1993). In notched specimens failure initiation is assumed to take place when σ_{eq}, calculated from Equation (6), has reached a maximum. In other words it is assumed that the weakest link theory applies to this type of specimen geometry.

Experiments were performed on specimens with the same geometries as those simulated. The results obtained with these specimens are shown in Figure 16, where they can be compared with those derived from the model. In this case the results corresponding to probabilities of 0.10, 0.50 and 0.90 are reported. A good correspondence between the experimental and simulated data is obtained for the notched specimens, i.e. for stress triaxiality ratios of ≈ 0.60, ≈ 0.80 and ≈ 1.20, corresponding to notched specimens. However, the values of the ductility for the smooth tensile specimens predicted from the simulation are below those which were measured. This arises from a size effect, which is investigated below. It was assumed that the smooth tensile specimens were uniformly deformed, but this hypothesis is not appropriate for these large-grain materials. Observation of the surface of the tensile specimens showed that the deformation was preferentially localized in certain grains which were favourably oriented.

The size effect was investigated both experimentally and numerically, using the same type of specimen geometries but with a minimum diameter of 6 or 15 mm instead of 10 mm. The notch radii were modified accordingly, in order to keep the same shape for the specimens and thus the same stress triaxiality ratio. The results are reported in Figure 17, where the calculated or measured values for the strains to failure are normalized by those determined on the specimens with a minimum diameter of 6 mm. The experimental results show that increasing the size of the specimens produces a reduction in the average ductility and a decrease in the scatter, as expected. The results of the numerical simulation are quite consistent with these observations.

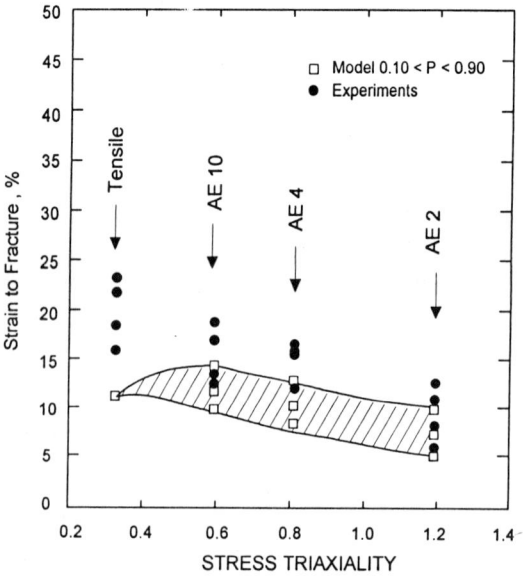

Fig. 16 Variation of the strain to failure with stress triaxiality ratio. Experimental results obtained on smooth and on notched specimens - Comparison with model.

3.4.2. Coupled modelling

The model and the numerical simulation presented above are based on a strong assumption indicated by Equation (8). The Gurson-Tvergaard potential of Equation (6) predicts that theoretically failure occurs when the volume fraction of cavities reaches a critical value $= 1/q_1$ (≈ 0.66). For this value of f the load-bearing capacity of the material is reduced to zero. However, it is well known that in most materials, failure occurs for mean values of the volume fraction of cavities much lower than 0.66 (see e.g.. Pineau, 1992). This is the main reason why an accelerating function for the evolution of the volume fraction of cavities was introduced (see Tvergaard and Needleman, 1984). Otherwise the Gurson-Tvergaard potential would largely overpredict the ductility of the materials. The main reason for this discrepancy lies in the fact that locally the volume fraction of cavities can be much larger than the mean value. It is therefore necessary to consider the effect of porosity distribution on ductile fracture. In this respect the cast duplex stainless steels, which present strong inhomogeneities, as illustrated previously, provide a good example. The effect of a non-uniform distribution of porosity on flow localization and failure in a porous material has already been investigated by Becker (1987). This author showed numerically that the ductility was largely reduced when a non-uniform distribution of cavities was introduced in the model.

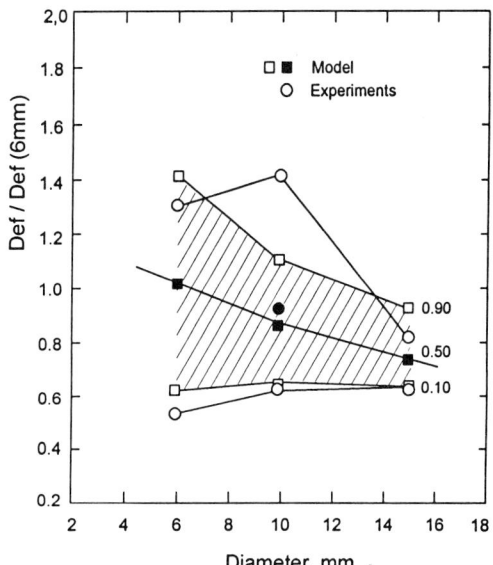

Fig. 17 Size effect on ductility. Strain to failure normalized by that corresponding to a 6 mm diameter specimen as a function of the diameter - Experiments and calculations.

This effect of the distribution of cavities in duplex stainless steels has been modelled recently in the author's laboratory (Devillers-Guerville et al, 1994). In the following we present these preliminary results, which are based only on 2D calculations but which illustrate the potentialities offered by a fully coupled model.

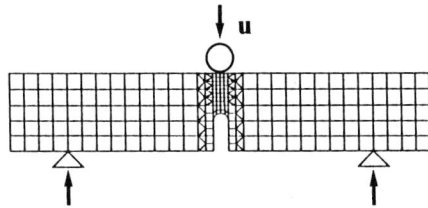

Fig. 18 Finite element meshes for a 3 point bend specimen.

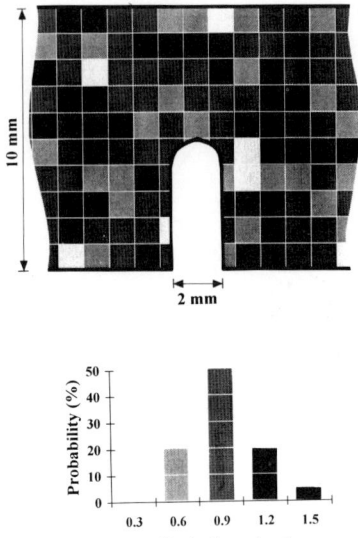
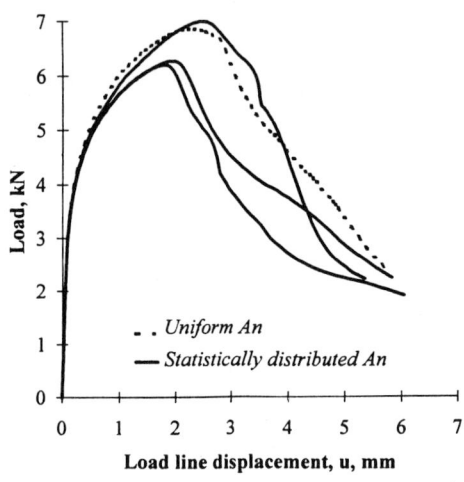

Fig. 19 a) Cells of various nucleation rates, A_n distributed in a 3P bend specimen; b) Distribution function of A_n.

Fig. 20 Load-Load line displacement curves calculated for 3P bend specimen with a uniform value of A_n and statistically distributed values.

The Gurson-Tvergaard potential was used to simulate the fracture toughness of Charpy U specimens. The meshes used to simulate the mechanical behaviour of these 3-point bend specimens are shown in Figure 18. A Monte Carlo type simulation similar to that used in the preceding section was applied in order to distribute the clusters with various nucleation rates (Figure 19a). A simplified distribution function for A_n was used (Figure 19b). The specimens were assumed to be deformed under plane strain conditions. The clusters with various values of A_n were assumed to be formed by cylinders parallel to the notch front with a square shape in a cross-section of the specimens. This assumption concerning the shape of the clusters was made in order to model the specimens with a 2D calculation. The numerical computations were this time made with Equation (6), which has been implemented in the finite element code (Besson, 1994). Figure 20 compares the load-load line displacement curves obtained with a uniform distribution of A_n with those obtained with several numerical simulations corresponding to different sampling of A_n. It is observed that in all cases the softening effect due to the presence of porosity, which is included in the Gurson-Tvergaard potential, is able to reproduce the fact that the loading curves reach a maximum. The softening effect is usually more pronounced in the simulations involving non-uniform distribution of the nucleation rate. This corresponds to the expected situation, since the probability of finding local values of A_n larger than the mean value will produce an acceleration in crack growth. Figure 21 shows the evolution of the calculated volume fraction of cavities in a numerical simulation with a random distribution of A_n.

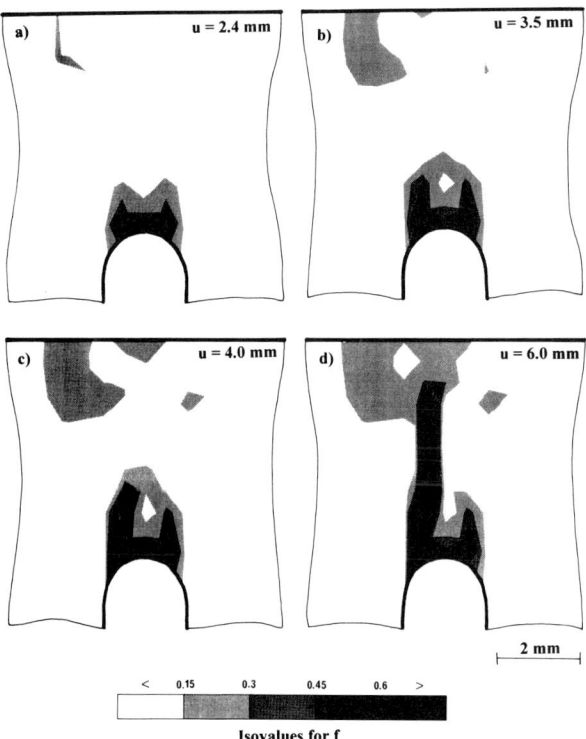

Fig. 21 Numerical simulation of a 3P bend specimen with statistically distributed values of An. Contours showing the position of isovalues of the porosity, f as a function of the load-line displacement.

The material breaks when the local volume fraction of cavities is larger than 0.66. The contours giving the isovalues of f are shown in Figure 21. The evolution of the contour giving the isovalues of f larger than ≈ 0.60 therefore indicates the position of the crack which is initiated and then propagates through the remaining ligament of the specimen. The main advantage of this type of fully coupled model lies in the fact that it is thus possible to simulate not only crack initiation but also crack growth. However, this advantage is counter-balanced by the difficulties in the calculations, not only those associated with convergence problems. In particular a realistic simulation of the non-uniform distribution of the clusters in cast duplex stainless steels, as in other materials, would require 3D calculations in which the grains would be distributed not only in the plane of the specimens but also through their thickness.

4. Conclusions

1. Homogenization techniques have proved to be extremely useful in order to predict the overall behaviour of polycrystals or multiphase materials from the knowledge of the elementary behaviour of single crystals or single phases. In particular it has been possible in a number of circumstances to include the effect of non-uniform distribution of local strains and stresses on the mechanical response of crystalline materials which occurs at various scales within it, including the intergranular (or interphase) scale and the transgranular scale. The effect of crystallographic texture on the anisotropy in plastic deformation can also be predicted. Rapid developments in computers are enabling us to deal with increasingly complex problems. However, much progress has to be made in at least two areas : (i) Better categorization is required regarding the problems strictly relevant to "microscopic" theories and those for which the micro-macro" approaches based on the extension of mixture laws can be applied; (ii) Micro-macro approaches would be more powerful if they could include geometrical effects.

2. Considerable progress needs to be made in predicting the damage behaviour of the materials, with more attention given to the proper treatment of the various scales involved. The mechanics of porous materials has enabled us to make significant progress in the modelling of ductile fracture, as illustrated by the specific case of cast duplex stainless steels. In this field we are becoming better able to categorize the effect of microstructure and to recognize in particular the importance of the statistical variations which are either present initially in the materials or develop progressively during their deformation.

3. The main interest of fully coupled damage models, in which the damage evolution during deformation contributes to the work softening of the material, lies in the fact that they can easily be used to model all stages of fracture, including crack initiation and crack growth. However, in a fully coupled model such as that proposed by Gurson-Tvergaard-Needleman (G.T.N.), it is necessary to take into account the non-uniformity in the damage distribution, otherwise a somewhat arbitrary accelerating function has to be introduced describing the variation in the volume fraction of porosity with deformation. In the numerical application of the G.T.N. model to aged cast duplex stainless steels in which continuous cavity nucleation plays a predominant role, it is possible to avoid the introduction of this arbitrary function by introducing the non-uniformity in the distribution of cavity nucleation rate.

Acknowledgements

The author would like to acknowledge many fruitful discussions with J. Besson, G. Cailletaud and Ph. Pilvin from the Centre des Matériaux and all his former and present PhD students, in particular, P. Joly, Ph. Achon, S. Jeunehomme and L. Devillers-Guerville.

References

Achon, P. (1964), *Comportement et ténacité d'alliages d'aluminium à haute résistance*. PhD Thesis, Ecole des Mines, Paris.

Argon, A.S., Im, J. and Safoglu, R. (1975), Cavity formation from inclusions in ductile fracture. *Met. Trans.*, **6A**, 825.

Asaro, R.J. (1975), Elastic plastic theory and kinematic-type hardening, *Acta Metall.*, **23**, 1255.

Ashby, M.F. (1971), The deformation of plastically nonhomogeneous alloys. In : *Strengthening methods in crystals*, A. Kelly and R.B. Nicholson, Eds, Elsevier, Amsterdam, Ch.3, p.137.

Becker, R. (1987), The effect of porosity distribution on ductile failure. *J. Mech. Phys. Solids*, **35**, 577.

Beremin, F.M. (1981), Cavity formation from inclusions in ductile fracture of A508 steel. *Met. Trans.*, **12A**, 723.

Berg, C.A. (1969), Plastic dilatation and void interaction. In : *Inelastic behavior of solids*, Kaninen M.K. et al., Eds, McGraw Hill, p.171.

Berveiller, M. and Zaoui, A. (1979), An extension of the self-consistent scheme to plastically flowing polycrystals, *J. Mech. Phys. Solids*, **6**, 326.

Besson, J. (1994), *Zebulon Code - Porous Potential class*. Internal Report JB1. Centre des Matériaux - Ecole des Mines, Paris.

Bonnet, S., Bourgoin, J., Champredonde, J., Guttmann, D., and Guttmann, M. (1990), Relationship between evolution of mechanical properties of various cast duplex stainless steels and metallurgical parameters. *Materials Science and Engineering*, **6**, 221.

Cailletaud, G. (1988), Une approche micromécanique du comportement des polycristaux. *Revue Phys. Appl.*, **23**, 353.

Cailletaud, G. (1992), A micromechanical approach to inelastic behaviour of metals. *Int. J. of Plasticity*, **8**, 55.

Cottrell, A.H. (1989), Strength of grain boundaries in pure metals. *Mat. Science and Technology*, **5**, 1165.

Cottrell, A.H. (1990a), Strenthening of grain boundaries by segregated interstiels in iron. *Mat. Science and Technology*, **6**, 121.

Cottrell, A.H. (1990b), Strength of grain boundaries in impure metals. *Mat. Science and Technology*, **6**, 325.

Devillers-Guerville, L., Besson, J., Pineau, A., and Eripret, C. (1994), Rupture ductile d'éprouvettes KCU et KCV d'aciers austéno-ferritiques fragilisés par vieillissement. *Revue de Métallurgie*, **9**, 1334.

Eshelby, J.D. (1957), The determination of the elastic field of an ellipsoïdal inclusion, and related problems. *Proc. Roy. Soc. London*, **A241**, 376.

Gologanu, M., Leblond, J.B., and Devaux, J. (1994), Approximate models for ductile metals containing nonspherical voids - Case of axisymmetric oblate ellipsoïdal cavities. *Trans. ASME*, **116**, 290.

Gurson, A.L. (1977), Continuum theory of ductile rupture by void nucleation and growth. Part I : Yield criteria and flow rules for porous ductile media. *J. of Engineering Materials and Technology*, **99**, p.2.

Hill, R. (1950), *The mathematical theory of plasticity*. Clarendon Press.

Hill, R. (1965), Continuum micromechanics of elastoplastic polycrystals. *J. Mech. Phys. Solids*, **13**, 89.

Jeunehomme, S, (1991). *Relations propriétés mécaniques-microstructure des aciers de structure perlitique*. PhD Thesis, Ecole des Mines, Paris.

Joly, P., Cozar, R., and Pineau, A. (1990), Effect of crystallographic orientation of austenite on the formation of cleavage cracks in ferrite in an aged duplex stainless steel. *Scripta Metall.*, **24**, 2235.

Joly, P., (1992), *Etude de la rupture d'aciers inoxydables austéno-ferritiques moulés, fragilisés par vieillissement à 400 °C*, Ph.D. Thesis, Ecole des Mines, Paris.

Joly, P., Meyzaud, Y., and Pineau, A. (1992), Micromechanims of fracture of an aged duplex stainless steel containing a brittle and a ductile phase : Development of a local criterion of fracture. In : *Advances in local fracture/damage models for the analysis of engineering problems*, J.H. Giovanola and A.J. Rosakis, Eds, ASME AMD, vol.137, p.151.

Joly, P., Pineau, A., and Meyzaud, Y. (1993)., Fracture micromechanisms of an aged duplex stainless steel. Application to the simulation of the fracture of notched tensile and compact tension specimens. *International Seminar on Micromechanics of Materials*, Fontainebleau, France - 6-8 July.

Kantidis, E., Marini, B., and Pineau, A. (1994a), A criterion for intergranular brittle fracture of a low alloy steel. *Fatigue Fract. Engng Mater. Struct.*, **17**, 619.

Kantidis, E., Marini, B., Allais, L., and Pineau, A. (1994b), Validation of a statistical criterion for intergranular brittle fracture of a low alloy steel through uniaxial and biaxial (tension-torsion) tests. *Int. Journal of Fracture*, **66**, 273.

Kröner, E. (1958), Berechnung der elastischen Konstanten des Vielkristalls aus den Konstanten des Einkristalls, *Z. Physik*, **151**, 504.

Kröner, .E. (1961), Zur plastichen Verformumg des Veilkristalls, *Acta Metall.*, **9**, 155.

Lemaitre, J. and Chaboche, J.L. (1985). *Mécanique des matériaux solides*, Dunod.

Lieurade, H.P. (1987), *Rôle de la microstructure sur la résistance à l'écaillage d'un acier à rail.* IRSID Report PE 4611, November.

Lin, T.H. (1957), Analysis of elastic and plastic strains of a face-centred cubic crystal. *J. Mech. Phys. Solids*, **5**, 143.

Louchet, F. (1992), Private communication.

Mecamat (1993), *International seminar on micromechanics of materials.* Moret sur Loing, France, 6-8 July. Eyrolles, Paris.

Meyzaud, Y., Ould, P., Balladon, P., Bethmont, M., and Soulat, P. (1988), Tearing resistance of aged cast austenitic stainless steel. *Proceedings NUC SAFE 88*, Avignon, France, p. 193.

Mudry, F. (1982), *Etude de la rupture ductile et de la rupture par clivage d'aciers faiblement alliés.* PhD. Thesis. Université de Technologie de Compiègne, France.

Perrin, G., and Leblond, J.B., (1990), Analytical study of a hollow sphere made of plastic porous material and subjected to hydrostatic tension - Application to some problems in ductile fracture of metals. *Int. J. of Plasticity*, **6**, 677.

Pilvin, P. (1993), Private communication.

Pineau, A., and Joly, P. (1991), Local versus global approaches to elastic-plastic fracture mechanics; Application to ferritic steels and a cast duplex stainless steel. In: *Defect Assessment in Components - Fundamentals and Applications*, J.G. Blauel and K.H. Schwalbe. ESIS/EGF Publication 9. MEP London, p.381.

Pineau, A. (1992), Global and local approaches to fracture - Transferability of laboratory tests results to components. In : *Topics in fracture and fatigue*, A.S. Argon. Ed. Springer Verlag. Ch.6, p.197.

Rousselier, G. (1981), Finite deformation constitutive relations including ductile fracture damage. In: *Three dimensional constitutive relations and ductile fracture*, Nemat-Nasser Ed., North Holland, Amsterdam, p.331.

Rousselier, G. (1987), Ductile fracture models and their potential in local approach of fracture. *Nuclear Engineering and Design*, **105**, 97.

Sunwoo, M., Fine, M.E., Meshi, M., and Stone, D.H. (1982), Cyclic deformation of pearlitic eutectoid rail steel. *Metall. Trans.*, **13A**, 2035.

Taylor, G.I. (1938), Plastic strain in metals, *J. Inst. Metals*, **62**, 307.

Tvergaard, V. (1981). Influence of voids on shear band instabilities under plane strain conditions. *Int. J. of Fracture*, **17**, 389.

Tvergaard, V., and Needleman, A. (1984), Analysis of the cup-cone fracture in a round tensile bar. *Acta Metall.*, **32**, 157.

Yamamoto, H. (1978), Conditions for shear localization in the ductile fracture of void containing materials, *Int. J. of Fracture*, **14**, 347.

Viggo Tvergaard*

Micromechanics of Damage in Metals

Reference: Tvergaard, V. (1995), Micromechanics of Damage in Metals. In: *Mechanical Behaviour of Materials* (ed. A. Bakker), Delft University Press, Delft, The Netherlands, pp. 23-43.

Abstract: Much understanding of failure mechanisms in metals has been gained by analyses of a characteristic unit cell, and the results have been used to develop constitutive relations. Such studies are here illustrated by results for ductile fracture at room temperature, where the nucleation and growth of voids to coalescence is the dominant fracture mechanism. Results are also presented for creep damage at elevated temperatures, where cavitation occurs mainly at the grain boundaries, and where interaction between neighbouring cavities as well as interaction between neighbouring grains has to be accounted for, including mechanisms such as grain boundary sliding. Finally, studies relating to metal-ceramic systems are considered, including metal-matrix composites, ductile particle reinforced ceramics, and ceramics bonded to thin ductile layers.

1. Introduction

Failure in metal alloys occurs by a number of different mechanisms, including brittle cleavage fracture at low temperatures or high strain rates, and fatigue failure under cyclic loading. In the present paper the focus will be on ductile failure in the room temperature range, on creep rupture at elevated temperatures, and on failure in metal-ceramic systems. The discussion is based on a number of micromechanical studies of the different failure mechanisms, with the main interest centred on the behaviour under monotonic loading.

In the range where metals undergo significant plastic straining prior to failure, the growth of small voids to coalescence with neighbouring voids is the dominant fracture mechanism. The voids nucleate mainly at second phase particles, by decohesion of the particle-matrix interface or by particle fracture, and subsequently the voids grow due to large plastic straining of the surrounding material. Early studies focussed on the growth of a single void in an infinite solid (McClintock,1968; Rice and Tracey,1969), while later on most studies considered solids with a certain volume fraction of voids, e.g. the porous ductile material model of Gurson (1977) and various improved versions of this model. The discussion here will include recent full three-dimensional cell-model studies for a material containing a periodic array of spherical voids, studies of the effect of a kinematic hardening porous material model, and results of an elastic-viscoplastic nonlocal model for a porous ductile material, in which an integral condition is used to introduce the effect of a characteristic material length scale.

During creep at elevated temperatures damage occurs mainly by the nucleation and growth of cavities on grain boundary facets perpendicular to the maximum principal tensile stress (Cocks and Ashby,1982; Argon,1982). Micromechanical studies of grain boundary cavity growth by the combined influence of diffusion and dislocation creep (Needleman and Rice, 1980) combined with a model for cavitation on a full facet (Rice, 1981) and expressions for the creep rate of a microcracked solid (Hutchinson, 1983) were used to propose a set of constitutive relations for creep in polycrystals with grain boundary cavitation (Tvergaard, 1984b). These constitutive relations are presented briefly here. Furthermore, some recent results of a more elaborate multi-grain cell-model (Van der Giessen

---

* *Department of Solid Mechanics, Technical University of Denmark, DK-2800 Lyngby, Denmark*

and Tvergaard, 1994a) are presented which make it possible to estimate the time to final material failure by the link-up of neighbouring facet microcracks.

In materials containing mixtures of metals and ceramics the mechanical behaviour is affected by the fact that plastic flow is highly constrained by the presence of elastic material. Furthermore, failure at metal-ceramic interfaces is an important damage mechanism. The behaviour of such material systems is illustrated by cell-model results for the damage of metals reinforced by short ceramic fibres. In addition the toughening effect of ductile particles in a brittle ceramic is demonstrated by micromechanical analyses of a particle bridging a brittle matrix crack. For the case of two ceramics joined by a thin metal layer, recent results are given for the effect of plasticity on interface crack growth.

2. Void Growth and Ductile Fracture

Studies of void growth in ductile metals have attracted much attention, as void coalescence is an important mechanism of ductile fracture. Early estimates of the critical strain for coalescence were based on micromechanical studies of the growth of a single void in an infinite elastic-plastic solid (McClintock, 1968; Rice and Tracey, 1969). However, most subsequent ductile fracture studies have focussed on approximate elastic-plastic material descriptions that account for the average effect of void nucleation and growth. The most widely known dilatant plasticity model for the average response of a ductile porous solid is that developed by Gurson (1977), based on approximate upper-bound solutions for a rigid perfectly plastic spherical unit cell containing a concentric spherical void.

Gurson's (1977) approximate yield condition is of the form $\Phi(\sigma^{ij}, \sigma_M, f) = 0$, where f is the void volume fraction, σ^{ij} is the average macroscopic Cauchy stress tensor, and σ_M is an equivalent tensile flow stress representing the actual microscopic stress state in the matrix material. To improve the agreement with accurate numerical micromechanical studies for small values of f Tvergaard (1981, 1982) suggested modifying parameters q_1 and q_2, and Tvergaard and Needleman (1984) proposed a function $f^*(f)$ for better representation of void coalescence. The modified Gurson yield condition takes the form

$$\Phi = \frac{\sigma_e^2}{\sigma_M^2} + 2q_1 f^* \cosh\left(\frac{q_2}{2} \frac{\sigma_k^k}{\sigma_M}\right) - 1 - \left(q_1 f^*\right)^2 = 0 \tag{2.1}$$

where $\sigma_e = \sqrt{3 s_{ij} s^{ij} / 2}$ is the macroscopic effective Mises stress, and $s^{ij} = \sigma^{ij} - G^{ij} \sigma_k^k / 3$ is the stress deviator. For $q_1 = q_2 = 1$ and $f^* = f$ Equation (2.1) is Gurson's yield condition, and for $f = 0$ Equation (2.1) reduces to the standard Mises yield condition. In the modified Gurson model $q_1 > 1$ is used (e.g. $q_1 = 1.5$), and the function $f^*(f)$ is taken to be

$$f^*(f) = \begin{cases} f & , \text{ for } f \leq f_C \\ f_C + \dfrac{f_U^* - f_C}{f_F - f_C}(f - f_C) & , \text{ for } f > f_C \end{cases} \tag{2.2}$$

Here f_F denotes the void volume fraction at final fracture, so that $f^*(f_F) = f_U^* = 1/q_1$, while f_C is a critical void volume fraction beyond which the modification according to Equation (2.2) is applied. A number of studies have been carried out using the values $f_C = 0.15$ and $f_F = 0.25$.

In the presentation of equations a Lagrangian formulation of the field equations is employed here, with a material point identified by the coordinates x^i in the reference configuration. The metric tensors in the current configuration and the reference configuration are denoted G_{ij} and g_{ij} respectively, with determinants G and g, and $\eta_{ij} = \frac{1}{2}(G_{ij} - g_{ij})$ is the Lagrangian strain tensor. The contravariant components of the Cauchy stress tensor σ^{ij} and the Kirchhoff stress tensor τ^{ij} are related by the expression $\tau^{ij} = \sqrt{G/g}\, \sigma^{ij}$.

A detailed micromechanical study of the predictions obtained by the modified Gurson model, Equations (2.1)-(2.2), have been carried out by Koplik and Needleman (1988), using an axisymmetric unit cell-model containing a single spherical void. These accurate numerical model studies include both the early stages, where overall plastic yielding and void growth take place, and the transition into the final stage, where only local plastic flow takes place in the ligaments between neighbouring voids. A fixed ratio of the transverse and axial macroscopic true stresses was prescribed in these cell-model analyses, considering different levels of stress triaxiality and non-hardening as well as hardening material behaviour. Rather good agreement between stress-strain curves and void growth vs. strain curves was found both in these studies and in similar studies by Becker et al. (1988), provided that the value f_C is chosen as a function of the initial void volume fraction, f_I, ranging from $f_C = 0.03$ at $f_I = 0.0104$ to $f_C = 0.12$ at $f_I = 0.06$.

A different type of model for ductile fracture was proposed by Thomason (1985, 1990), based on approximate upper bound solutions for non-hardening rigid-plastic solids. Some of these solutions rely on plane strain analyses, while others rely on three-dimensional solutions, in which ellipsoidal voids are represented by square-prismatic voids. In these studies the discussion mainly focussed on the transition from a weak dilatational yield surface (such as that proposed by Gurson, 1977) to a strong dilatational yield surface representing a stage where void coalescence develops. The type of behaviour predicted by these simple upper-bound solutions was studied by Richelsen and Tvergaard (1993, 1994), using detailed numerical unit cell analyses and comparisons with dilatant plasticity predictions.

Both plane strain analyses and full three-dimensional analyses were carried out by Richelsen and Tvergaard (1993, 1994), based on non-hardening elastic-plastic material behaviour or on the corresponding elastic-viscoplastic material model. The accurate numerical plane strain analyses were used to study the transition into a ligament necking mode. It has been shown that transition predictions obtained by simple upper-bound solutions are in some cases quite reasonable. The full 3D cell-model analyses were carried out for a material containing a periodic array of spherical voids (Figure 1), subject to fixed ratios of the average principal stresses, so that

$$S_2 = \kappa S_1 \quad , \quad S_3 = \gamma S_1 \quad (2.3)$$

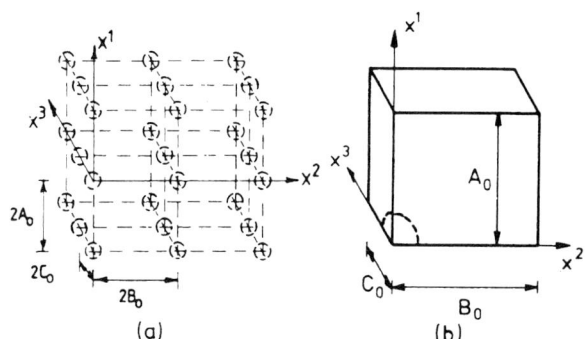

Fig. 1 (a) Model material containing a periodic array of spherical voids.
(b) 3D cell model.

where κ and γ are prescribed constants. For $\kappa = \gamma$ a good approximation of the material behaviour can be obtained by an axisymmetric analysis, as has been shown by Hom and McMeeking (1989) and Worswick and Pick (1990) by comparison with 3D analyses. However, for $\kappa \neq \gamma$ the full 3D analysis is necessary. Several analyses with many different values of κ and γ, were carried out for the initial void volume fraction $f_I = 0.04$, with values of the stress triaxiality parameter, S_m/S_e, ranging from 0.15 to 3. In each case the peak stresses were compared with the prediction corresponding to the approximate yield condition of Equation (2.1), and it was found (Figure 2) that the agreement is generally rather good for $q_1 = 1.5$ in Equation (2.1). The 3D computations were continued to larger strains, beyond the onset of failure by necking in the ligament between neighbouring voids, at least for the higher stress triaxialities investigated. There are also cases where the voids grow towards coalescence without reaching an obvious ligament necking mode. In Equation (2.2), for $f_I = 0.04$, the parameter values $f_C = 0.094$ and $f_F = 0.25$ were used, and comparison with the 3D numerical results showed that Equation (2.2) gives a useful representation of final failure.

In relation to Thomason's (1985, 1990) criticism of the Gurson model, the main thing to note is that computations such as those of Koplik and Needleman (1988) for non-hardening materials subject to axisymmetric conditions, or the 3D computations of Richelsen et al. (1994), account for exactly the same mode of final failure by localized necking of the intervoid matrix as that analysed in the approximate 3D rigid-plastic upper-bound solutions of Thomason (1985). The only difference is that the numerical solutions are much more accurate and do not require approximations of spherical or ellipsoidal voids as square-prismatic voids. Thus, on the basis of the comparisons with the modified Gurson model it may be concluded that by using the function $f^*(f)$ in Equation (2.1), with appropriate values of q_1, f_C and f_F, this model gives a rather good description of the ductile failure process, including both the initial growth of voids and the final failure by coalescence.

In real materials voids are not uniformly distributed, as was assumed in the unit cell-model studies mentioned above. The effect of porosity distribution was examined experimentally by

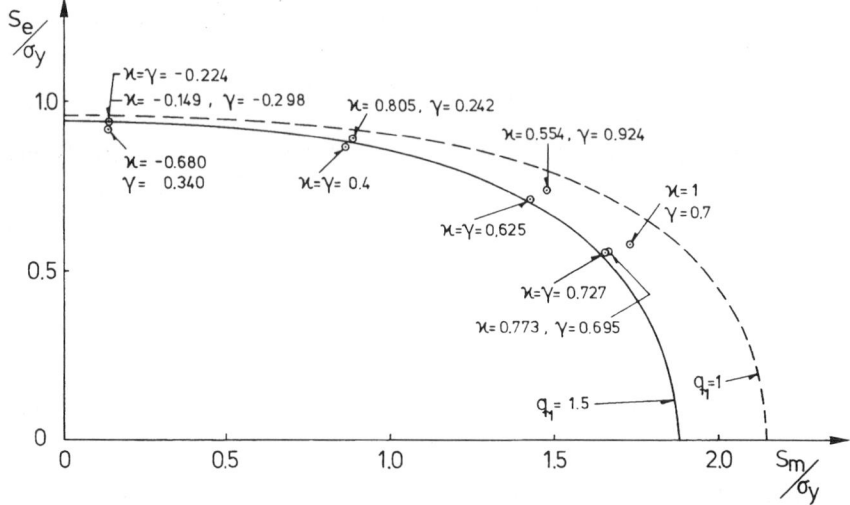

Fig. 2 Maximum stress points for initial void volume fraction 0.04, obtained by 3D cell model analyses for non-hardening material. Comparison with approximate yield conditions (2.1); from Richelsen and Tvergaard (1994).

Magnusen et al. (1988) who drilled arrays of holes in sheet tensile test specimens and compared the test results for specimens with different hole distributions but with a fixed area fraction of holes. In these specimens final failure by void coalescence involved sheet necking in the ligaments between neighbouring holes. Recent plane stress simulations of full specimens with many holes (Becker and Smelser, 1994) were able to reproduce the fracture path observed in the experiments. These investigations show that earlier onset of failure by void coalescence is the most significant effect of a random void distribution. In addition, the non-uniform void distribution also results in a slightly lower strength prior to the onset of failure. Such strength reductions prior to localization were also found by Needleman and Kushner (1990) for plane strain conditions, in a model material with larger voids represented in terms of a Gurson material by regions of high void volume fraction.

Most of the studies of the effect of porosity on plastic flow have focussed on spherical voids; but there has recently been increasing interest in studying the influence of deviations from spherical void geometry. Lee and Mear (1992) studied an ellipsoidal void in an infinite plastic or viscous medium. Gologanu, Leblond and Devaux (1993) worked on developing approximate plastic yield criteria for a material containing a certain volume fraction of ellipsoidal voids. Ponte Castañeda and Zaidman (1994) have also considered the effect of non-spherical voids and have proposed a constitutive model, including evolution equations for appropriate state variables such as the void volume fraction and the aspect ratios of the typical void.

Predictions of plastic flow localization are very sensitive to whether or not a vertex forms on the yield surface (Rice, 1977). Kinematic hardening can be used to model approximately the effect of a rounded vertex, and it has been found both for necking in biaxially stretched shells (Tvergaard, 1978) and for shear localization under plane strain conditions (Hutchinson and Tvergaard, 1981) that kinematic hardening predictions are rather similar to those found for a solid that develops a sharp vertex on the yield surface. In order to incorporate this effect in studies of ductile fracture, Mear and Hutchinson (1985) have suggested a kinematic hardening model for a porous ductile material, and Tvergaard (1987a) has extended this model to account for void nucleation. In fact, a family of isotropic/kinematic hardening yield surfaces are introduced by taking the radius, σ_F, of the yield surface for the matrix material to be given by

$$\sigma_F = (1-b)\sigma_y + b\,\sigma_M \tag{2.4}$$

where σ_y and σ_M are the initial yield stress and the matrix flow stress respectively, and the parameter b is a constant in the range $[0,1]$. The constitutive relations are formulated such that for $b=1$ they reduce to Gurson's (1977) isotropic hardening model, whereas a pure kinematic hardening model appears for $b=0$. For $b=1$ and $f \equiv 0$ the expressions reduce to J_2 flow theory.

Then, instead of Equation (2.1), the approximate yield condition for the porous solid is taken to be of the form

$$\Phi = \frac{\tilde{\sigma}_e^2}{\sigma_F^2} + 2q_1 f^* \cosh\left(\frac{q_2\,\tilde{\sigma}_k^k}{2\,\sigma_F}\right) - 1 - \left(q_1 f^*\right)^2 = 0 \tag{2.5}$$

where α^{ij} denotes the stress components at the centre of the yield surface, $\tilde{\sigma}^{ij} = \sigma^{ij} - \alpha^{ij}$, $\tilde{\sigma}_e = \sqrt{3\tilde{s}_{ij}\tilde{s}^{ij}/2}$ and $\tilde{s}^{ij} = \tilde{\sigma}^{ij} - G^{ij}\tilde{\sigma}_k^k/3$. For $q_1 = 1$ Equation (2.5) is the expression proposed by Mear and Hutchinson (1985), which coincides with that of Gurson (1977) for $b=1$. The plastic part of the macroscopic strain increment, $\dot{\eta}_{ij}^P$, and the effective plastic strain increment, $\dot{\varepsilon}_M^P$, for the matrix material are taken to be related by

$$\tilde{\sigma}^{ij} \dot{\eta}_{ij}^P = (1-f)\sigma_F \dot{\varepsilon}_M^P \qquad (2.6)$$

Using the uniaxial true stress natural strain curve for the matrix material, $\dot{\varepsilon}_M^P = (1/E_t - 1/E)\dot{\sigma}_M$, an expression for the matrix flow stress increment, $\dot{\sigma}_M$, is obtained from Equation (2.6). Furthermore, the change in the void volume fraction, $\dot{f}$, during an increment of deformation is still taken to consist of a contribution from the growth of existing voids and a contribution from the nucleation of new voids.

Unrealistic oscillatory stress predictions have been found for kinematic hardening solids subject to large shear strains; it has been shown (Dafalias, 1983; Loret, 1983), however, that these stress oscillations disappear if certain corotational stress rates other than the Jaumann rate are used in the finite strain generalization of the constitutive law. For the ductile porous material model Tvergaard and Van der Giessen (1991) incorporated alternative stress rates involving corotation with the crystal substructure spin (the elastic spin) rather than with the continuum spin. The Jaumann rate $\overset{\triangledown}{\sigma}^{ij}$ of the Cauchy stress and the alternative rate $\overset{\circ}{\sigma}^{ij}$ are defined by

$$\overset{\triangledown}{\sigma}^{ij} = \dot{\sigma}^{ij} + \left(G^{ik}\sigma^{jl} + G^{jk}\sigma^{il}\right)\dot{\eta}_{kl} \qquad (2.7)$$

$$\overset{\circ}{\sigma}^{ij} = \overset{\triangledown}{\sigma}^{ij} + \left(G_{ik}\sigma^{jl} - \sigma^{ik}G^{jl}\right)\omega_{kl}^P \qquad (2.8)$$

$$\omega_{ij}^P = \frac{1}{2}\rho\left(G^{ik}\dot{\eta}_{lj}^P - \dot{\eta}_{ik}^P G_{lj}\right)\alpha^{kl} \qquad (2.9)$$

In a macroscopic plasticity theory the separation of continuum spin in an elastic part and a plastic part is not defined, and Equation (2.9) is an assumed constitutive law for the plastic spin, in which the factor ρ appears as an additional material function.

The plastic part of the strain rate is taken to be (Tvergaard and Van der Giessen, 1991)

$$\dot{\eta}_{ij}^P = \frac{1}{H}m_{ij}^G m_{kl}^F \overset{\circ}{\sigma}^{kl} \qquad (2.10)$$

where the expressions for H and the tensors m_{ij}^G and m_{ij}^F are given by Tvergaard (1987a). The hardening rule, expressing the evolution of the yield surface centre during a plastic increment, is taken to be

$$\overset{\circ}{\alpha}^{kl} = \dot{\mu}\,\tilde{\sigma}^{kl}\;,\quad \dot{\mu} \geq 0 \qquad (2.11)$$

where the value of the parameter $\dot{\mu}$ is determined by the consistency condition $\dot{\Phi} = 0$.

The effect of kinematic hardening on the onset of localization in a shear band is illustrated in Figure 3. For an initial imperfection in the form of void nucleating particles, $\Delta f_N = 0.01$, inside the band, the minima in Figure 3a give the first critical strain for localization. It is seen that higher yield surface curvature ($b < 1$) does indeed promote early onset of shear band formation; but it is also noted that values of ρ in Equation (2.9) chosen large enough to avoid unrealistic oscillatory stress predictions under pure shear (i.e. the dashed and dotted curves in Figure 3) result in a small delay in the predicted localization.

The different versions of a porous ductile material model discussed so far are all standard local constitutive relations, which do not represent a material length scale. It is well known that if no length scale is included in the problem formulation, the softening material behaviour typical of situations near final failure will tend to give localized damage in regions as narrow as possible within the mesh resolution. Thus numerical predictions of final fracture will show an inherent mesh sensitivity. The relevant length scale may be introduced by directly specifying the size and spacing of larger voids; it has been found (Needleman and Tvergaard, 1994) that this can remove the inherent mesh dependence. Alternatively, a set of nonlocal constitutive relations may be applied.

Pijaudier-Cabot and Bažant (1987) and Barenblatt (1992) suggested nonlocal constitutive formulations in which delocalization relates to the damage mechanism, and the same idea has been used by Leblond, Perrin and Devaux (1993) to propose a nonlocal version of the Gurson model. An elastic-viscoplastic version of this nonlocal porous ductile material model, recently applied by Tvergaard and Needleman (1994), will be briefly introduced here.

For the elastic-viscoplastic material the modified Gurson yield condition according to Equation (2.1) is applied as a flow potential (Pan et al., 1983), so that the plastic part of the strain-rate is given by

$$\dot{\eta}_{ij}^P = \Lambda \frac{\partial \Phi}{\partial \sigma^{ij}} \quad , \quad \Lambda = \frac{(1-f)\sigma_M \dot{\varepsilon}_M^P}{\sigma^{kl} \frac{\partial \Phi}{\partial \sigma^{kl}}} \quad (2.12)$$

where the expression for Λ is found from the isotropic hardening version of Equation (2.6). The matrix plastic strain rate is taken to be given by

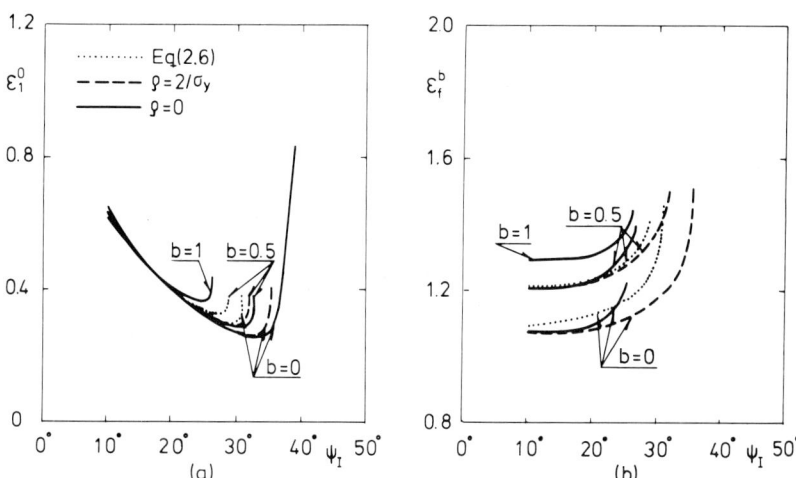

Fig. 3 Shear band formation in kinematic hardening porous solid (2.5) under uniaxial plane strain tension. Band with initial orientation Ψ_I contains small initial imperfection. (a) Localization strains; (b) Fracture strains inside band; from Tvergaard and Van der Giessen (1991).

$$\dot{\varepsilon}_M^P = \dot{\varepsilon}_0 \left(\frac{\sigma_M}{g(\varepsilon_M^P)} \right)^{1/m} , \quad g(\varepsilon_M^P) = \sigma_0 \left(1 + \frac{\varepsilon_M^P}{\varepsilon_0} \right)^N \qquad (2.13)$$

where $\dot{\varepsilon}_0$ is a reference strain-rate, m and N are the strain-rate hardening exponents and σ_0 and ε_0 are reference stress and strain quantities, respectively. Furthermore, with attention restricted to strain-controlled nucleation, the local growth of existing voids and the local nucleation of new voids is given by

$$\dot{f}_{local} = (1-f) G^{ij} \dot{\eta}_{ij}^P + \mathcal{D}\,\dot{\varepsilon}_M^P \qquad (2.14)$$

Void nucleation, specified in terms of $\mathcal{D}(\varepsilon_M^P)$, is taken here to follow a normal distribution (Chu and Needleman, 1980).

The rate of increase in the void volume fraction in the material point at location x^i in the reference configuration is obtained from the local values in Equation (2.14) by (Leblond et al., 1993; Pijauder-Cabot and Bažant, 1987)

$$\dot{f}(x^i) = \frac{1}{W(x^i)} \int_V \dot{f}_{local}(\hat{x}^i) w(x^i - \hat{x}^i) d\hat{V} \qquad (2.15)$$

where V is the volume of the body in the reference configuration. In Equation (2.15) Tvergaard and Needleman (1994) used the weight function specified by

$$w(z^i) = \left[\frac{1}{1+(z/L)^p} \right]^q , \quad W(x^i) = \int_V w(x^i - \hat{x}^i) d\hat{V} \qquad (2.16)$$

Here, $z = \sqrt{G_{ij} z^i z^j}$ and L is a positive material characteristic length. The two remaining parameter values were chosen as $p = 8$ and $q = 2$. The local formulation corresponds to the limit $L \to 0$. With $L > 0$, $\dot{f}(x^i) \equiv \dot{f}_{local}$ when $\dot{f}_{local}$ is spatially uniform. Hence, nonlocality is associated with spatial gradients in $\dot{f}$. It is noted that the function w is chosen such that $w > 0$ for $z < L$ and $w \approx 0$ for $z > L$, with a relatively narrow transition region; $w = 0.25$ at $z/L = 1.0$, $w = 0.021$ at $z/L = 1.25$ and $w = 0.0014$ at $z/L = 1.5$. In computations a cut-off length $L_c = 1.5L$ may be used such that w in Equation (2.15) is taken to be zero for $z > L_c$.

There is no direct micromechanical basis for the nonlocal constitutive description resulting from Equation (2.15), but it is assumed that the material characteristic length L is of the order of magnitude of the average void spacing. Two different planar model problems were used by Tvergaard and Needleman (1994) to study the effect of the delocalization, a material that develops plastic flow localization in shear bands and a metal matrix composite. Figure 4 shows contours of constant void volume fraction for the material, with a doubly periodic array of soft spots that trigger the development of a periodic pattern of crossing shear bands. Due to symmetries only a rectangular subregion with initial dimensions A_0 and B_0 needs to be analysed numerically. The figure shows results at the average logarithmic strain $\varepsilon_I = 0.520$ for four different meshes in a material subject to uniaxial plane strain tension. The material length scale is specified by $L/B_0 = 0.1$ and it is seen that due to the nonlocal formulation the shear band width is rather mesh-insensitive, whereas

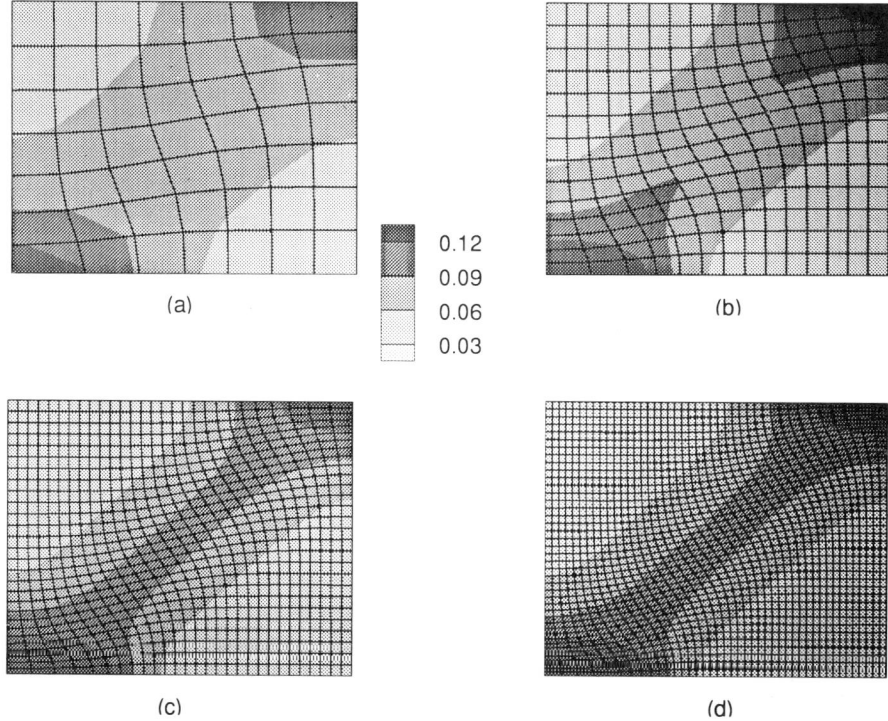

Fig. 4 Contours of constant volume fraction in non-local material at $\varepsilon_I = 0.520$, for material length $L = 0.1\ B_0$; from Tvergaard and Needleman (1994).

computations for $L = 0$ would show a band width corresponding to one quadrilateral element in all four cases. For the planar model of a metal matrix composite it was found (Tvergaard and Needleman, 1994) that the delocalization described by Equation (2.15) removes the strong mesh sensitivity resulting from high gradients of the field quantities near sharp fibre corners.

3. Models for Creep Rupture

In polycrystalline metals at elevated temperatures creep rupture can occur by a number of different mechanisms. Ashby and Dyson (1984) distinguished four broad categories of such mechanisms, one of which involves the nucleation and subsequent growth of microscopic cavities, leading to fracture by coalescence. Another category involves failure by the loss of the cross-sectional area associated with large strains, for example in the form of necking, while the third category relates to degradation of the microstructure by thermal coarsening of particles or by a dislocation substructure induced acceleration of creep. The fourth category covers damage by gas-environmental attack.

Based on a number of micromechanical cell-model studies a set of constitutive relations for creep in polycrystals with grain boundary cavitation was developed by Tvergaard (1984b), as an extension of work by Rice (1981) and Hutchinson (1983). This material model describes dislocation creep of the grains in terms of power-law creep, incorporates the effect of grain boundary sliding, and accounts for the nucleation and growth of cavities occurring mainly on grain boundary facets perpendicular to the maximum principal tensile stress (Hull and Rimmer, 1959; Cocks and Ashby, 1982; Argon, 1982). The mixed influence of diffusion and dislocation creep on cavity growth is accounted for (Needleman and Rice, 1980; Sham and Needleman, 1983), as well as the mechanism of creep-constrained cavitation (Dyson, 1976). Failure due to loss of the cross-sectional area is also directly represented, since finite strains are accounted for, and an approximate representation of creep acceleration due to microstructure degradation was included (Tvergaard, 1987b). Thus the material model accounts for the first three categories of creep damage mechanisms discussed by Ashby and Dyson (1984).

Inside the grains, where dislocation creep is modelled in terms of power-law creep, the creep part of the Lagrangian strain rate is

$$\dot{\eta}_{ij}^C = \dot{\varepsilon}_e^C \frac{3}{2} \frac{s_{ij}}{\sigma_e} \quad , \quad \dot{\varepsilon}_e^C = \dot{\varepsilon}_0 \left(\frac{\sigma_e}{\sigma_0} \right)^n \tag{3.1}$$

where $\dot{\varepsilon}_0$ and σ_0 are reference strain rate and stress quantities, and n is the creep exponent. The effective Mises stress is $\sigma_e = \sqrt{3 s_{ij} s^{ij}/2}$, with $s^{ij} = \tau^{ij} - G^{ij} \tau_k^k / 3$.

Grain boundary cavities are assumed to be distributed over a facet, with average spacing $2b$ and radius a, and the diffusion along the void surface is assumed to be sufficiently rapid, relative to the diffusion along the grain boundary, to maintain the quasi-equilibrium spherical caps void shape (see Figure 5a). For the angle ψ a value around 70° is typical. The growth of a single void in the

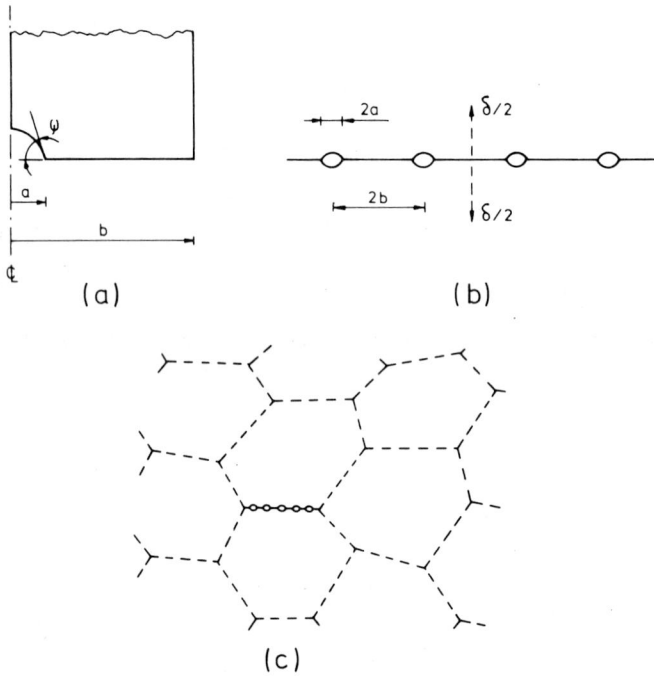

Fig. 5 (a) Spherical-caps shape of a single cavity
(b) Equally spaced cavities on a grain boundary
(c) An isolated, cavitated grain boundary facet in a polycrystalline material.

spherical caps shape, by combined grain boundary diffusion and dislocation creep, was studied numerically by Needleman and Rice (1980) and Sham and Needleman (1983). At sufficiently low tensile stresses, cavity growth by grain boundary diffusion is dominant. Then, the rate of growth of the cavity volume is obtained by the rigid grains model, analysed early on by Hull and Rimmer (1959) and subsequently modified by various authors, including Needleman and Rice (1980), who found

$$\dot{V}_1 = 4\pi \mathcal{D} \frac{\sigma_n - (1-f)\sigma_s}{\ln(1/f) - (3-f)(1-f)/2} \qquad (3.2)$$

Here σ_s is the sintering stress, f is the area fraction of the grain boundary which is cavitated, and $\mathcal{D} = D_B \delta_B \Omega / kT$ is the grain boundary diffusion parameter, where $D_B \delta_B$ is the boundary diffusivity, Ω is the atomic volume, k is Boltzmann's constant, and T is the absolute temperature.

At sufficiently high stresses power-law creep growth of the cavities would dominate, and the corresponding volumetric growth rate $\dot{V}_2$ can be approximated, based on results of Budiansky, Hutchinson and Slutsky (1982). Using this, Sham and Needleman (1983) suggested the following expression for the rate of growth of the cavity volume at high stress triaxialities

$$\dot{V} = \dot{V}_1 + \dot{V}_2, \text{ for } \frac{a}{L} \leq 10, f = \max\left\{\left(\frac{a}{b}\right)^2, \left(\frac{a}{a+1.5L}\right)^2\right\} \qquad (3.3)$$

where the stress and temperature-dependent length scale of the cavitation process, $L - (\mathcal{D}\sigma_e / \dot{\varepsilon}_e^C)^{1/3}$, was introduced by Needleman and Rice (1980). Equation (3.3) shows good agreement with numerically determined growth rates, both for high and low triaxialities (Needleman and Rice, 1980; Tvergaard, 1984a). From Equation (3.3) the rate of growth of the cavity radius is found as $\dot{a} = \dot{V} / (4\pi a^2 h(\psi))$. In addition to these expressions for cavity growth, an approximate way of incorporating continuous nucleation of new cavities in the material model was proposed by Tvergaard (1984b), based on nucleation observations discussed by Argon (1982) and Dyson (1983).

In order to describe the complex mechanism of grain boundary cavitation in more detail, Rice (1981) suggested that an isolated, cavitating grain boundary facet can be modelled as a penny-shaped crack, as illustrated in Figure 5c. The average rate of separation, $\dot{\delta}$, of the grains adjacent to a facet (Figure 5b) is given as the average rate of opening of the crack, using an expression obtained by He and Hutchinson (1981) modified to account for a non-zero normal tensile stress, σ_n, on the crack surfaces (Tvergaard, 1984a)

$$\dot{\delta} = \beta^* \frac{S^* - \sigma_n}{\sigma_e} \dot{\varepsilon}_e^C 2R \qquad (3.4)$$

Here R is the current radius of the crack, β^* is a constant, S^* is the value that the normal stress on the facet would have if there was no cavitation and $\dot{\varepsilon}_e^C$ is the effective creep strain rate according to Equation (3.1). Equation (3.4) is based on the assumption that the facet is normal to the maximum principal tensile stress. If V is the cavity volume, and $2b$ is the average cavity spacing (see Figure 5), the average separation across the facet is $\delta = V / (\pi b^2)$, and thus the rate of separation is also given by (Tvergaard, 1984b)

$$\dot{\delta} = \frac{\dot{V}}{\pi b^2} - \frac{2V}{\pi b^2} \frac{\dot{b}}{b} \qquad (3.5)$$

The requirement that Equations (3.4) and (3.5) are equal determines the value of the normal stress, σ_n, and the cavity growth rate, $\dot{V}$, (which is a function of σ_n according to Equations (3.2) and (3.3)). This model describes the phenomenon of creep-constrained cavitation (Dyson, 1976), when diffusion is so rapid relative to dislocation creep of the adjacent grains that cavity growth occurs for $\sigma_n / S^* \approx 0$ (Rice, 1981; Tvergaard, 1984a).

For a material containing many cavitating grain boundary facets an expression for the macroscopic creep strain rates used by Tvergaard (1984b) is based on that derived by Hutchinson (1983) for a solid containing a certain density of penny-shaped microcracks, modified to account for a non-zero stress, σ_n, on the crack surfaces. With further modifications to approximately describe grain boundary sliding (Tvergaard, 1985) the expression takes the form

$$\dot{\eta}_{ij}^C = \dot{\varepsilon}_e^C \left[\frac{3}{2} \frac{s_{ij}}{\sigma_e} (f^*)^n + \rho^* \left\{ \frac{3}{2} \frac{n-1}{n+1} \frac{s_{ij}}{\sigma_e} \left(\frac{S^* - \sigma_n}{\sigma_e} \right)^2 + \frac{2}{n+1} \frac{S^* - \sigma_n}{\sigma_e} m_{ij}^* \right\} \right] \quad (3.6)$$

If there is no sliding, the value of f^* is unity, and ρ^* and m_{ij}^* reduce to special values ρ and $m_{ij} = \bar{n}_i \bar{n}_j$ respectively, where $\bar{n}_i$ is the facet normal in the current configuration, and ρ reflects the density of cavitating facets. The expressions for ρ and β and approximations used to represent grain boundary sliding are given by Tvergaard (1985a). The material model based on Equations (3.1)-(3.6) has been used in finite element computations to predict the creep rupture behaviour of structural components, for example in a number of analyses of creep crack growth (Tvergaard, 1986, 1990c; Li et al., 1988). An example of predicted damage distributions around a crack tip is shown in Figure 6 at two stages, with damage measured by a/b and completely failed elements in black.

Micromechanical models of creep rupture, including those mentioned above, have focussed on the time to cavity coalescence on a characteristic grain boundary facet, and this time has been used as an estimate of the lifetime. However, final intergranular failure occurs as microcracks on

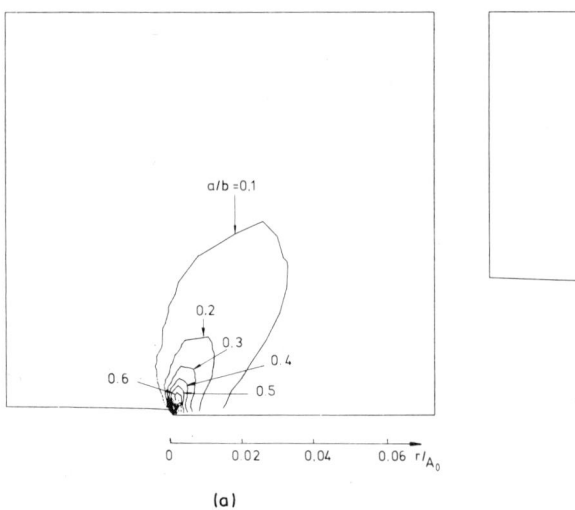
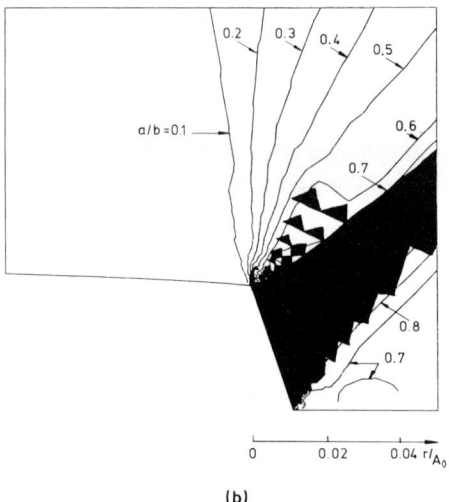

Fig. 6 Predicted creep damage distribution around crack tip in double edge cracked panal, for $a_I/L_N = 0.1$ and continuous cavity nucleation;
(a) At $t/t_f^0 = 0.064$. (b) At $t/t_f^0 = 0.686$; from Tvergaard (1990c).

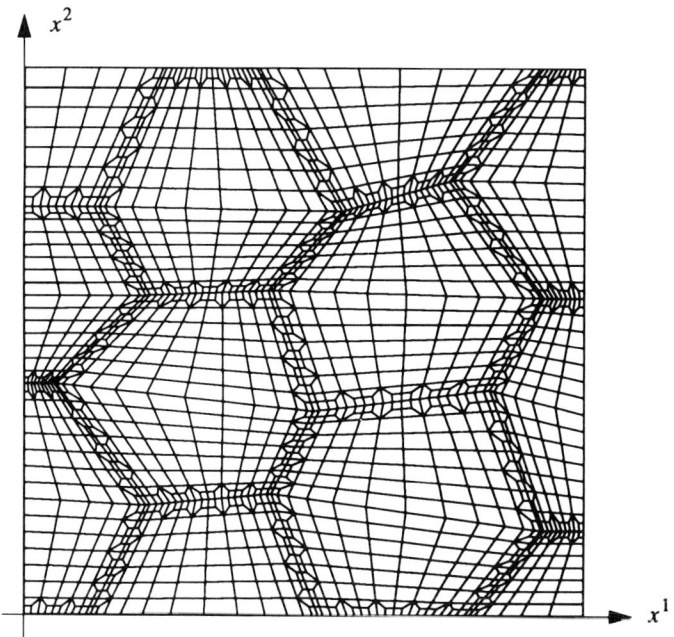

Fig. 7 Finite element mesh used in analysis of a planar multigrain cell model, with cavitation and sliding at all grain boundaries; from Van der Giessen and Tvergaard (1994c)

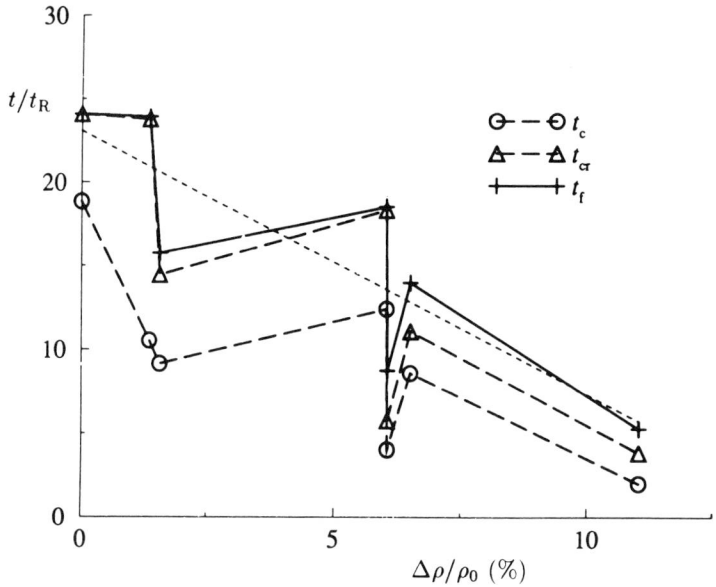

Fig. 8 Times to first coalescence, t_c/t_R, times to first microcrack, t_{cr}/t_R, and times to final creep failure, t_f/t_R, for planar multigrain cell model with free grain boundary sliding and $a_I/L_I = 0.025$. Increasing $\Delta \rho$ marks increasing randomness of aggregate; from Van der Giessen and Tvergaard (1994c).

neighbouring grain boundary facets link-up, and studying this process requires more elaborate micromechanical models. The deformation of grains in a real three-dimensional polycrystal is much more constrained by the deformations of neighbouring grains than is found in a planar model. An axisymmetric model problem (Tvergaard, 1985; Van der Giessen and Tvergaard, 1991) captures the main effect of this 3D constraint and is still relatively simple, but this model is not able to fully represent the interaction with neighbouring cavitating facets leading to link-up of facet microcracks. Detailed analyses for full 3D aggregates containing many grains are not yet numerically feasible, but some insight into the linking-up process can be obtained from planar aggregates.

A plane strain model of a polycrystalline aggregate was investigated by Van der Giessen and Tvergaard (1994a, 1994b). Here cavitation is accounted for on all grain boundary facets, using Equations (3.2), (3.3) and (3.5) to approximately represent the influence of diffusion and dislocation creep on cavity growth as well as the process of continuous cavity nucleation. Furthermore, grain boundary sliding following a linear viscous relationship between shear stress and sliding rate (Ashby, 1972) is incorporated. Thus, depending on the values of material parameters and stress level, the model represents behaviour ranging from free grain boundary sliding to no sliding at all. In addition, the cavitation model makes it possible to represent different cavity densities or different nucleation rates on different grain boundary facets, so that cavity coalescence leading to an open facet microcrack will tend to occur earlier at some facets than at others. It is noted that Hsia, Parks and Argon (1991) used a rather similar planar multigrain model with free sliding to analyse the effect of microcrack density on the rate of opening of a facet microcrack.

A number of different analyses for the planar multigrain model have shown that typically about 20 to 30% of the lifetime remains after the first occurrence of an open facet microcrack. In a polycrystal built up of identical hexagonal grains (Van der Giessen and Tvergaard, 1994b) this of course depends on differences between nucleation rates on different facets. The effect of random variations in microstructure has also been studied (Van der Giessen and Tvergaard, 1994c) by analysing planar grain aggregates such as that illustrated by the finite element mesh in Figure 7. The results of computations for six different random microstructures are summarized in Figure 8, where increasing values of $\Delta\rho$ are a measure of increasing randomness in facet widths. The grain boundaries slide freely, the nucleation parameters are the same for all facets, and the material is subject to overall uniaxial plane strain tension. Figure 8 shows the three landmarks in the lifetime of the material: the time, t_c, to first cavity coalescence on any facet, the time, t_{cr}, to the first formation of a full facet microcrack, and the final time to full loss of integrity, t_f.

4. Metal-Ceramic Systems

Materials containing mixtures of metals and ceramics, e.g. metal-matrix composites, ductile particle reinforced ceramics, or ductile layer reinforced ceramics, have emerged as a class of materials capable of various advanced structural applications. The mechanics of these materials differs from more traditional materials in a number of ways, including the fact that here plastic flow tends to occur under highly constrained conditions and that failure at the interface between different phases plays an important role.

In the case of metal-matrix composites, the reinforcement by ceramic fibres increases the stiffness and the tensile strength compared with that of the metal alloy, but the reinforcement also leads to reduced ductility and fracture toughness due to the limited strength of fibres and the fibre-matrix interface. Thus observations of matching pairs of fracture surfaces for Al-SiC short-fibre composites show the occurrence of failure by debonding as well as by brittle fracture of elongated SiC particles aligned with the tensile direction (Zok et al., 1988; Lagacé and Lloyd, 1989; Mummery and Derby, 1991). A parametric understanding of the effect of different material parameters, such as the shape and distribution of fibres, the volume fraction, the strength of the interface and fibres, the matrix yield stress and strain hardening, can be obtained by numerical cell-model analyses, which allow for an accurate representation of the stress and strain fields, including local stress peaks at sharp fibre edges.

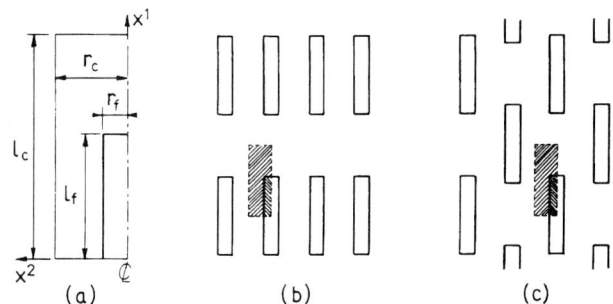

Fig. 9 Periodic arrays of parallel short fibres.
(a) Axisymmetric cell model.
(b) Transversely aligned fibres.
(c) Transversely staggered fibres.

For a metal reinforced by a periodic array of parallel, short fibres Tvergaard (1993, 1994a) focussed on analysing the effect of fibre-matrix decohesion and fibre fracture. Such materials can be modelled in terms of the axisymmetric cell-model shown in Figure 9a. With appropriate boundary conditions this unit cell represents a material with transversely aligned fibres, as illustrated in Figure 9b; for different boundary conditions the same unit cell can also be used to approximate a material with transversely staggered fibres (Figure 9c), as suggested by Tvergaard (1990a). The axisymmetric unit cell was used by Nutt and Needleman (1987) and Christman et al. (1989) to analyse composites with transversely aligned short fibres. This type of axisymmetric cell-model can be considered a good approximation of a three-dimensional array of hexagonal cylinders (Tvergaard, 1982); but a square array is also well approximated. It is noted that in addition to fibre breakage or decohesion, failure within the matrix material alone has also been observed in metal matrix composites (see, for example, Needleman et al., 1993).

Debonding at the matrix-fibre interface is modelled by a cohesive zone formulation (Tvergaard, 1990b), which is a generalization of that proposed by Needleman (1987) and applied by Nutt and Needleman (1987) for composites. In the generalized model a set of interface constitutive relations gives the dependence of the normal and tangential tractions, T_n and T_t, on the corresponding components, u_n and u_t, of the displacement difference across the interface, and the model is chosen such that in pure normal separation ($u_t \equiv 0$) it coincides with that of Needleman (1987). A non-dimensional parameter λ is defined as

$$\lambda = \sqrt{\left(\frac{u_n}{\delta_n}\right)^2 + \left(\frac{u_t}{\delta_t}\right)^2} \qquad (4.1)$$

and a function $F(\lambda)$ is chosen as

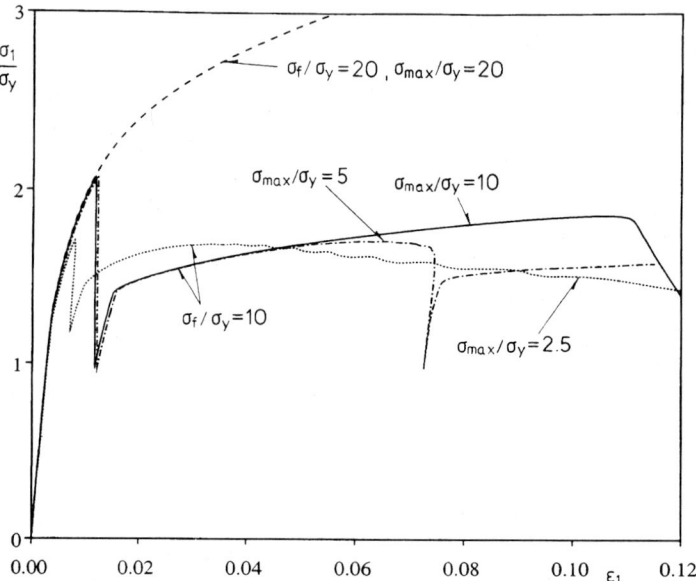

Fig. 10 Average stress-strain curves for metal matrix composite with transversely aligned short fibres, under uniaxial tension. Fibre volume fraction is 0.13; from Tvergaard (1990a).

$$F(\lambda) = \frac{27}{4}\sigma_{max}\left(1 - 2\lambda + \lambda^2\right), \text{ for } 0 \le \lambda \le 1 \tag{4.2}$$

Then, as long as λ is monotonically increasing, the interface tractions are taken to be given by the expressions

$$T_n = \frac{u_n}{\delta_n}F(\lambda) \quad , \quad T_t = \alpha\frac{u_t}{\delta_t}F(\lambda) \tag{4.3}$$

In pure normal separation ($u_t \equiv 0$) the maximum traction is σ_{max}, total separation occurs at $u_n = \delta_n$, and the work of separation per unit interface area is $9\sigma_{max}\delta_n/16$. In pure tangential separation ($u_n \equiv 0$) the maximum traction is $\alpha\sigma_{max}$, total separation occurs at $u_t = \delta_t$, and the work of separation per unit interface area is $9\alpha\sigma_{max}\delta_t/16$. The values of the four parameters δ_n, δ_t, σ_{max} and α have to be chosen such that the maximum traction and work of separation in different situations are well approximated.

For SiC reinforced aluminium the fibre elastic modulus is much higher than that for aluminium ($E_f \approx 5.7E_{Al}$), and the fibres are approximated as rigid in the analyses. The fibre fracture criterion employed is specified as a critical value of the average tensile stress on a cross-section. A recent study of particulate fracture in MMC's, by Finot et al. (1993) has shown that a critical average tensile stress gives a good approximation of predictions obtained by assuming an initial crack inside a particulate.

Figure 10 shows examples of predicted stress-strain curves for a power hardening matrix material with $\sigma_y/E = 0.005$ and $n = 7.66$ and for a fibre volume fraction $f = 0.13$, with fibre aspect ratio 5. Here the material is subject to overall uniaxial tension in the fibre direction, the fibres are transversely aligned (as in Figure 9b), and different values of the fibre strength, σ_f, and the interface strength, σ_{max}, are considered. Computations for the same sets of material parameters but

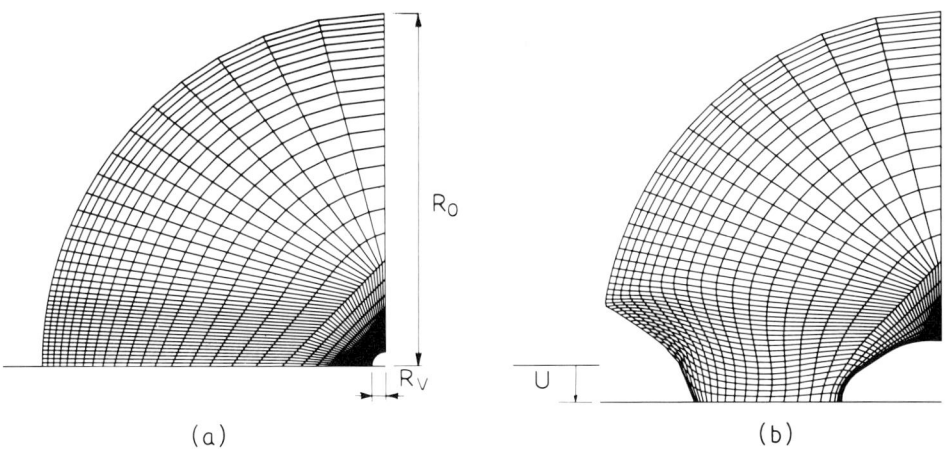

Fig. 11 Ductile particle bridging a brittle matrix crack. (a) Initial mesh in particle with small central void. (b) Deformed mesh at half crack opening U; from Tvergaard (1994b).

for transversely staggered fibres (Tvergaard, 1993), showed failure by breakage or debonding at lower values of the average stress, σ_1, and the average strain, ε_1. Thus, in a real material, where whiskers are more or less randomly distributed, it appears to be more likely that failure will initiate in regions where the fibres are transversely staggered than in regions with transversely aligned fibres.

The debonding model in Equations (4.1)-(4.3) has also been used to study a metal-ceramic system where the metal phase plays the opposite role, as a dispersed ductile phase added to a brittle ceramic in order to increase the fracture toughness. Here ductile particles intercepted by a brittle matrix crack undergo large elastic-plastic deformations during crack bridging, and experiments for Al_2O_3 reinforced by Al particles (Flinn, Rühle and Evans, 1989) have indicated that progressive particle debonding during bridging may significantly add to the toughening. Numerical model studies to investigate this type of behaviour were carried out by Tvergaard (1992), assuming that the ductile particles are spherical and that the particle centre is in the plane of the brittle matrix crack. It has been found that progressive particle-matrix debonding during bridging can significantly increase the additional fracture toughness obtained due to particle yielding.

The experimental observations of Flinn *et al.* (1989) showed that the failure of axisymmetric reinforcement zones often involves the nucleation of a single hole at the centre of the neck, which rapidly expands to final failure. This was analysed by Tvergaard (1994b) by assuming the initial presence of a small void in the particle centre. Figure 11 shows the initial mesh used for such a numerical study and a deformed mesh at a stage with the half crack opening U, where both debonding and large void growth are clearly visible. It is found in these computations that residual tensile stresses in the ductile particles, due to thermal contraction mismatch during cooling from the processing temperature, have a strong influence on the void growth. Thus the growth of a single void in a ductile particle may be strongly affected by the occurrence of a cavitation instability (Huang *et al.*, 1991; Tvergaard *et al.*, 1992).

In layered materials crack growth along an interface between a brittle ceramic and a ductile metal is also an important failure mechanism, with plastic yielding of the ductile phase adding significantly to the fracture toughness. In such interface crack growth the elastic solution governing the remote field has an oscillating crack tip singularity, so that the mode mixity (K_2 vs. K_1) varies with the distance from the crack tip. Both model experiments (Liechti and Chai, 1992; O'Dowd et al., 1992) and crack growth analyses (Tvergaard and Hutchinson, 1993) have shown that the crack grows most easily when the near tip fields are close to mode 1 conditions.

When two ceramics are joined by a thin metal layer, plastic deformations of the metal will also add to the fracture toughness of this sandwich system. It has been found by Tvergaard and Hutchinson (1994), using a cohesive zone model to study crack growth along one of the thin layer interfaces, that in addition to the stress ratio, σ_{max} / σ_y, the toughening effect of the layer depends strongly on the ratio h / R_0 (where h is the layer thickness). If h exceeds the reference plastic zone size, R_0, a significant toughening can be obtained when σ_{max} / σ_y is sufficiently large; for small values of h / R_0 the plastic deformations of the layer have no effect on the toughness. The results of Tvergaard and Hutchinson (1994) were obtained under the assumption of equal elastic moduli of the layers, but studies considering more realistic ratios of the elastic properties are now being carried out.

References

Argon, A.S. (1982), Mechanisms and mechanics of fracture in creeping alloys. In: *Recent Advances in Creep and Fracture of Engineering Materials and Structures* (eds. B. Wilshire and D.R.J. Owen), Pineridge, Swansea, pp. 1-52.

Ashby, M.F. (1972), Boundary defects, and atomistic aspects of boundary sliding and diffusional creep, *Surface Science* **31**, 498-542.

Ashby, M.F. and Dyson, B.F. (1984), *Creep damage mechanics and micromechanisms*, National Physical Laboratory, Report DMA(A), 77.

Barenblatt, G.I. (1992), Micromechanics of fracture, *Proc. Int. Congr. Theor. Appl. Mech.*, Haifa.

Becker, R. and Smelser, R.E. (1994), Simulation of strain localization and fracture between holes in an aluminum, *J. Mech. Phys. Solids* **42**, 773-796.

Becker, R., Needleman, A., Richmond, O. and Tvergaard, V. (1988), Void growth and failure in notched bars, *J. Mech. Phys. Solids* **36**, 317-351.

Budiansky, B., Hutchinson, J.W. and Slutsky, S. (1982), Void growth and collapse in viscous solids. In: *Mechanics of Solids. The Rodney Hill 60th Anniversary Volume* (eds. H.G. Hopkins and M.J. Sewell), Pergamon Press, Oxford, pp. 13-45.

Christman, T., Needleman, A., Nutt, S., and Suresh, S. (1989), On microstructural evolution and micromechanical modelling of deformation of a whisker-reinforced metal-matrix composite, *Materials Science and Engineering A* **107**, 49-61.

Chu, C.C. and Needleman, A. (1980), Void nucleation effects in biaxially stretched sheets, *J. Eng. Materials Technol.* **102**, 249-256.

Cocks, A.C.F. and Ashby, M.F. (1982), On creep fracture by void growth, *Progress in Materials Science* **27**, 189-244.

Dafalias, Y.F. (1983), Corotational rates for kinematic hardening at large plastic deformations, *J. Appl. Mech.* **50**, 561-565.

Dyson, B.F. (1976), Constraints on diffusional cavity growth rates, *Metal Science* **10**, 349-353.

Dyson, B.F. (1983), Continuous cavity nucleation and creep fracture, *Scripta Metallurgica* **17**, 31-37.

Finot, M., Shen, Y.-L., Needleman, A., and Suresh, S. (1993), *Micromechanical modelling of reinforcement fracture in particulate-reinforced metal-matrix composites*, Report, Division of Engineering, Brown University.

Flinn, B., Rühle, M., and Evans, A.G. (1989), Toughening in composites of Al_2O_3 reinforced with Al, Materials Department Report, University of California, Santa Barbara.

Gologanu, M., Leblond, J.-B. and Devaux, J. (1993), Approximate models for ductile metals containing non-spherical voids - Case of axisymmetric prolate ellipsoidal cavities, *J. Mech. Phys. Solids* **41**, 1723-1754.

Gurson, A.L. (1977), Continuum theory of ductile rupture by void nucleation and growth - I. Yield criteria and flow rules for porous ductile media, *J. Eng. Materials Technol.* **99**, 2-15.

He, M.Y. and Hutchinson, J.W. (1981), The penny-shaped crack and the plane strain crack in an infinite body of power-law material, *J. Appl. Mech.* **48**, 830-840.

Hom, C.L. and McMeeking, R.M. (1989), Void growth in elastic-plastic materials, *J. Appl. Mech.* **56**, 309.

Hsia, K.J., Parks, D.M. and Argon, A.S. (1991), Effects of grain boundary sliding on creep-constrained boundary cavitation and creep deformation, *Mech. Mater.* **11**, 43-62.

Huang, Y., Hutchinson, J.W. and Tvergaard, V. (1991), Cavitation instabilities in elastic-plastic solids, *J. Mech. Phys. Solids* **39**, 223-241.

Hull, D. and Rimmer, D.E. (1959), The growth of grain-boundary voids under stress, *Phil. Mag.* **4**, 673-687.

Hutchinson, J.W. (1983), Constitutive behaviour and crack tip fields for materials undergoing creep-constrained grain boundary cavitation, *Acta Metallurgica* **31**, 1079-1088.

Hutchinson, J.W. and Tvergaard, V. (1981), Shear band formation in plane strain, *Int. J. Solids Structures* **17**, 451-470.

Koplik, J. and Needleman, A. (1988), Void growth and coalescence in porous plastic solids, *Int. J. Solids Structures* **24**, 835.

Lagacé, H. and Lloyd, D.J. (1989), Microstructural analysis of Al-Sic composites, *Canadian Metallurgical Quarterly* **28**, 145-152.

Leblond, J.B., Perrin, G. and Devaux, J. (1993), Bifurcation effects in ductile metals with damage delocalization, *J. Appl. Mech.*, ASME, in press.

Lee, B.J. and Mear, M.E. (1992), Axisymmetric deformation of power-law solids containing a dilute concentration of aligned spheroidal voids, *J. Mech. Phys. Solids* **40**, 1805-1836.

Li, F.Z., Needleman, A. and Shih, C.F. (1988), Creep crack growth by grain boundary cavitation: crack tip fields and crack growth rates under transient conditions, *Int. J. Fracture* **38**, 241-273.

Liechti, K.M. and Chai, Y.-S. (1992), Asymmetric shielding in interfacial fracture under in-plane shear, *J. Appl. Mech.* **59**, 295-304.

Loret, B. (1983), On the effects of plastic rotation in the finite deformation of anisotropic elastoplastic materials, *Mechanics of Materials* **2**, 287-304.

Magnusen, P.E., Dubensky, E.M. and Koss, D.A. (1988), The effect of void arrays on void linking during ductile fracture, *Acta Metall.* **36**, 1503-1509.

McClintock, F.A. (1968), A criterion for ductile fracture by growth of holes, *J. Appl. Mech.* **35**, 363-371.

Mear, M.E. and Hutchinson, J.W. (1985), Influence of yield surface curvature on flow localization in dilatant plasticity, *Mechanics of Materials* **4**, 395-407.

Mummery, P. and Derby, B. (1991), The influence of microstructure on the fracture behaviour of particulate metal matrix composites, *Mater. Sci. Engng.* **A135**, 221-224.

Needleman, A. (1987), A continuum model for void nucleation by inclusion debonding, *J. Appl. Mech.* **54**, 525-531.

Needleman, A. and Kushner, A.S. (1990), An analysis of void distribution effects on plastic flow in porous solids, *Eur. J. Mech., A/Solids* **9**, 193-206.

Needleman, A. and Rice, J.R. (1980), Plastic creep flow effects in the diffusive cavitation of grain boundaries, *Acta Metallurgica* **28**, 1315-1332.

Needleman, A. and Tvergaard, V. (1994), Mesh effects in the analysis of dynamic ductile crack growth, *Engng. Fracture Mech.* **47**, 75-91.

Needleman, A., Nutt, S.R., Suresh, S. and Tvergaard, V. (1993), Matrix, reinforcement and interfacial failure. In *Fundamentals of Metal Matrix Composites* (eds. S. Suresh, A. Mortensen and A. Needleman), Butterworth-Heinemann, Boston, pp. 233-250,

Nutt, S.R. and Needleman, A. (1987), Void nucleation at fiber ends in Al-SiC composites, *Scripta Metallurgica* **21**, 705-710.

O'Dowd, N.P., Stout, M.G. and Shih, C.F. (1992), Fracture toughness of alumina/niobium interfaces: Experiments and analyses, *Phil. Mag. A* **66**, 1037.

Pan, J., Saje, M. and Needleman, A. (1983), Localization of deformation in rate sensitive porous plastic solids, *Int. J. Fracture* **21**, 261-278.

Pijaudier-Cabot, G. and Bažant, Z.P. (1987), Nonlocal damage theory, *J. Engng. Mech.*, ASCE **113**, 1512-1533.

Ponte Castañeda, P. and Zaidman, M. (1994), Constitutive models for porous materials with evolving microstructure, *J. Mech. Phys. Solids* **42**, 1459-1497.

Rice, J.R. (1977), The localization of plastic deformation. In: *Theoretical and Appl. Mech.* (ed. W.T. Koiter), North-Holland, pp. 207-220.

Rice, J.R. (1981), Constraints on the diffusive cavitation of isolated grain boundary facets in creeping polycrystals, *Acta Metallurgica* **29**, 675-681.

Rice, J.R. and Tracey, D.M. (1969), On the ductile enlargement of voids in triaxial stress fields, *J. Mech. Phys. Solids* **17**, 201-217.

Richelsen, A.B. and Tvergaard, V. (1993), Comparison of porous ductile material models and upper bound estimates. In: *Advanced Computational Methods for Material Modeling*, (eds. D.J. Benson *et al.*), ASME, AMD-Vol. 180, New York, pp. 33-39.

Richelsen, A.B. and Tvergaard, V. (1994), Dilatant plasticity or upper bound estimates for porous ductile solids, *Acta Metall. Mater.* **42**, 2561-2577.

Sham, T.-L. and Needleman, A. (1983), Effects of triaxial stressing on creep cavitation of grain boundaries, *Acta Metallurgica* **31**, 919-926.

Thomason, P.F. (1985), Three-dimensional models for the plastic limit-loads at incipient failure of the intervoid matrix in ductile porous solids, *Acta Metall.* **33**, 1079-1085.

Thomason, P.F. (1990), *Ductile fracture of metals*, Pergamon Press.

Tvergaard, V. (1978), Effect of kinematic hardening on localized necking in biaxially stretched sheets, *Int. J. Mech. Sci.* **20**, 651-658.

Tvergaard, V. (1981), Influence of voids on shear band instabilities under plane strain conditions, *Int. J. Fracture* **17**, 389-407.

Tvergaard, V. (1982), On localization in ductile materials containing spherical voids, *Int. J. Fracture* **18**, 237-252.

Tvergaard, V. (1984a), On the creep constrained diffusive cavitation of grain boundary facets, *J. Mech. Phys. Solids* **32**, 373-393.

Tvergaard, V. (1984b), Constitutive relations for creep in polycrystals with grain boundary cavitation, *Acta Metallurgica* **32**, 1977-1990.

Tvergaard, V. (1985), Effect of grain boundary sliding on creep constrained diffusive cavitation, *J. Mech. Phys. Solids* **33**, 447-469.

Tvergaard, V. (1986), Analysis of creep crack growth by grain boundary cavitation, *Int. J. Fracture* **31**, 183-209.

Tvergaard, V. (1987a), Effect of yield surface curvature and void nucleation on plastic flow localization, *J. Mech. Phys. Solids* **35**, 43-60.

Tvergaard, V. (1987b), Creep failure by degradation of the microstructure and grain boundary cavitation in a tensile test, *Acta Metallurgica* **35**, 923-933.

Tvergaard, V. (1990a), Analysis of tensile properties for a whisker-reinforced metal-matrix composite, *Acta Metall. Mater.* **38**, 185-194.

Tvergaard, V. (1990b), Effect of fibre debonding in a whisker-reinforced metal. *Materials Science and Engineering A* **125**, 203-213.

Tvergaard, V. (1990c), Effect of microstructure degradation on creep crack growth, *Int. J. Fracture* **42**, 145-155.

Tvergaard, V. (1992), Effect of ductile particle debonding during crack bridging in ceramics, *Int. J. Mech. Sci.* **34**, 635-649.

Tvergaard, V. (1993), Model studies of fibre breakage and debonding in a metal reinforced by short fibres, *J. Mech. Phys. Solids* **41**, 1309-1326.

Tvergaard, V. (1994a), *Fibre debonding and breakage in a whisker-reinforced metal*, Techn. Univ. of Denmark, DCAMM Report No. 479.

Tvergaard, V. (1994b), *Cavity growth in ductile particles bridging a brittle matrix crack*, Techn. Univ. of Denmark, DCAMM Report No. 491.

Tvergaard, V. and Van der Giessen, E. (1991), Effect of plastic spin on localization predictions for a porous ductile material, *J. Mech. Phys. Solids* **39**, 763-781.

Tvergaard, V. and Hutchinson, J.W. (1993), The influence of plasticity on mixed mode interface toughness, *J. Mech. Phys. Solids* **41**, 1119-1135.

Tvergaard, V., and Hutchinson, J.W. (1994), *Toughness of an interface along a thin ductile layer joining elastic solids*, Harvard University, Div. Appl. Sci., Report MECH-230.

Tvergaard, V. and Needleman, A. (1994), *Effects of nonlocal damage in porous plastic solids*, DCAMM Report No. 487.

Tvergaard, V., Huang, Y. and Hutchinson, J.W. (1992), Cavitation instabilities in a power hardening elastic-plastic solid, *Eur. J. Mech., A/Solids* **11**, 215-231.

Van der Giessen, E. and Tvergaard, V. (1991), A creep rupture model accounting for cavitation at sliding grain boundaries, *Int. J. Fracture* **48**, 153-178.

Van der Giessen, E. and Tvergaard, V. (1994a), Interaction of cavitating grain boundary facets in creeping polycrystals, *Mechanics of Materials* **17**, 47-69.

Van der Giessen, E. and Tvergaard, V. (1994b), Development of final creep failure in polycrystalline aggregates, *Acta Metall. Mater.* **42**, 959-973.

Van der Giessen, E. and Tvergaard, V. (1994c), Effect of random variations in microstructure on the development of final creep failure in polycrystalline aggregates, *Modelling Simul. Mater. Sci. Eng.* **2**, 721-738.

R. A. Schapery* and D. L. Sicking**

On Nonlinear Constitutive Equations for Elastic and Viscoelastic Composites with Growing Damage

Reference: Schapery, R.A. and Sicking, D.L. (1995), On Nonlinear Constitutive Equations for Elastic and Viscoelastic Composites with Growing Damage. In: *Mechanical Behaviour of Materials* (ed. A. Bakker), Delft University Press, Delft, The Netherlands, pp. 45-76.

Abstract: Homogenized constitutive equations for the mechanical behaviour of unidirectional fibre composites with growing damage are discussed. Emphasis is on resin matrices reinforced with high modulus, elastic fibre. Starting with time-independent behaviour, the thermodynamic foundation for a work potential-based model is reviewed. The model is then successfully applied to the characterization of a carbon/epoxy composite. Laminates with many different fibre-angle combinations are used to determine the basic material functions and to check the accuracy of the model for predicting response not used in the characterization process. Edge delamination of three different laminates is then predicted using the overall laminate work potential. Finally, methods of accounting for viscoelastic deformation effects and for rate-dependent damage evolution equations are described. It is shown that the time-independent theory can be readily extended to account for these complexities using approximations that are believed to be valid for many fibre composites.

1. Introduction

Structural composites such as continuous fibre-reinforced plastics are traditionally modelled as linear elastic materials. Although this may be acceptable for many engineering applications, it is well-known that the resin is highly nonlinear in shear and is viscoelastic. Even for apparently fibre-dominated response, such as when a unidirectional composite is under axial compression, the strength is directly dependent on the resin behaviour through a shear microbuckling phenomenon (Budiansky and Fleck, 1993). Damage growth in composites is also highly dependent on the resin behaviour. The most common forms of observed damage are broken fibres, delamination and transverse cracks (Friedrich, 1989); the latter are intraply cracks normal to the layer plane and parallel to the fibres. The effect of resin viscoelasticity on the time-dependent growth of transverse cracks is clearly seen in the experimental studies of Moore and Dillard (1990).

Here we shall be primarily concerned with homogenized constitutive equations for unidirectional layers which account for material nonlinearity without and with transverse cracks. The effect of the former source of nonlinearity without these cracks is illustrated in Figure 1 for the specific composite material which is studied in this paper. The difference in loading and unloading behaviour shows that the response is not elastic. Although this material exhibits viscoelastic behaviour, its effect at room temperature and for the strain rate used is negligible compared to the amount of hysteresis exhibited in Figure 1. The inelastic behaviour exhibited in this figure cannot be fully modelled using traditional plasticity theories because the unloading curves deviate considerably from a straight line with the initial Young's modulus E_0. Typically, for different angle-ply and unidirectional laminates, the residual strain upon load removal is only 20-40% of that for unloading along the E_0 line (Schapery, 1989a).

---

* *The University of Texas, Austin, TX 78712, USA*
** *The University of Nebraska at Lincoln, Lincoln, NE 68516, USA*

Hahn and Tsai (1973) introduced one of the earliest nonlinear material models for a unidirectional composite. They assume elastic behaviour and use a complementary strain energy density function that produces a third order relationship between shear stress and strain for shearing parallel to the fibres. For loading behaviour this model provides a fairly good representation of the shear behaviour up to at least moderate stress levels, but does not include nonlinearity between stress and strain normal to the fibres and coupling between normal and shearing variables. Most publications on nonlinear behaviour use incremental or deformation plasticity theory as a basis for developing constitutive equations with coupled normal and shear nonlinearities (e.g. Christensen, 1979, Sun and Chen, 1989 and Budiansky and Fleck, 1993). In the latter two references a three-parameter Ramberg-Osgood (power law) nonlinear representation is used, and unloading behaviour is not addressed. Sun and Chen (1989) show that experimental, uniaxial stress-strain behaviour of off-axis, unidirectional carbon/epoxy and boron/aluminium composites is characterized quite well by their model. Although their basic formulation is an incremental plasticity theory, the three in-plane stress components (referred to the principal material axes) are proportional to the applied axial load. As this is a case of proportional loading, the theory reduces to deformation theory and the strains may be derived from a complementary energy potential, just as if the material were elastic. The carbon/epoxy material studied in the present paper *cannot* be adequately represented by the Ramberg-Osgood equation (Schapery, 1995).

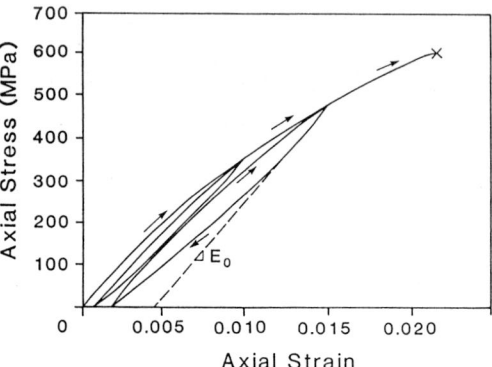

Fig. 1   Stress-strain behaviour of Hercules' AS4/3502 carbon/epoxy, angle-ply laminate [±30]$_{3S}$ showing inelastic behavior (after Schapery, 1989a).

Transverse cracking and delamination, which have been studied extensively, are characterized and predicted in most cases assuming the composite material is linearly elastic (e.g. Friedrich, 1989; Hashin, 1985; Masters and Reifsnider, 1982; O'Brien, 1982; Varna, 1992). The last reference has an extensive study of transverse cracks in crossply laminates, and consists of five papers on this subject as well as on local delamination. Nonlinear behaviour is taken into account in a few investigations such as that of Tsai et al. (1990) who account for the effect of fibre-induced elastic nonlinearity in a transverse cracking model, and that of Schapery et al. (1986) and Schapery (1989a) who introduce fibre- and matrix-induced nonlinear behaviour in deformation and delamination studies.

In structural applications of composite laminates, it may be necessary to account for one or more of the various types of nonlinear behaviour, damage and material time-dependence prior to the development of significant fibre fracture. Of course, which, if any, of the complexities that are important in an application depends on the loading, physical environment, material system and layup, etc. Schapery (1987, 1989a,b, 1990) developed an approach to material modelling that uses the same mathematical formalism for nonlinear elastic and inelastic behaviour and for damage evolution and its effect. It is based on thermodynamics with internal state variables, fracture mechanics and the experimental observation that the stresses and mechanical work for fibre-reinforced plastics and particle-reinforced rubber are often independent of many details of the deformation history (Lamborn and Schapery, 1988, 1993; Mast, et al., 1992; Schapery, 1987, 1991a). This limited path-dependence leads to the use of a work potential (or complementary work potential) which is analogous to strain energy (or complementary strain energy) for a nonlinear elastic material. The type of behaviour illustrated in Figure 1 may be combined with effects of transverse cracking and delamination using a work potential. Unlike classical incremental plasticity theory, this model uses total (rather than incremental) strain. The modelled behaviour may be independent of path (for a limited or wide range of paths) with non-proportional loading, and inelastic behaviour during loading is not necessarily tied to unloading behaviour and residual strains.

In this paper we shall use a work potential to account for nonlinear elastic behaviour of the fibres, inelastic behaviour of the matrix, transverse cracking and delamination. Only loading behaviour will be addressed here, apart from the foundational discussion in Section 2. A few examples of path-independence for unloading and reloading behaviour may be found in experimental investigations of axial-torsional behaviour of laminates (Lamborn and Schapery, 1988, 1993) and in theoretical studies based on a work potential (Schapery, 1989a,b).

Section 2 presents an overview of work potential theory. The reader is referred to Schapery (1990) for additional details and proofs. Development of specific time-independent constitutive equations using a work potential, and their experimental validation for a carbon/epoxy composite, are described in Section 3. Most of the material characterization is taken from Sicking's (1992) Ph.D. dissertation; however, there are also some new results here, especially on the transverse cracking problem. Finally, an extension of work potential theory to viscoelastic deformation and damage growth behaviour is outlined in Section 4. This extension is included here because it makes direct use of the time-independent theory and requires only a mathematically small modification of the time-independent constitutive equations.

2. Thermodynamic Foundation of the Work Potential

Generalized Notation and Basic Equations

The thermodynamically-based theory used here is first expressed in terms of generalized displacements q_j and their work-conjugate forces Q_j. This notation is employed as it provides freedom in using the basic theory for various applications, including unidirectional ply constitutive equations and overall laminate mechanical behaviour. For example q_j could be strain, displacement or rotation, and Q_j stress, force or moment. The material or structural system of interest may or may not have an initial unit volume, and full geometric nonlinearity is permitted in this section. For all processes of interest, the existence of a strain energy, W, is assumed, with the property that

$$Q_j = \partial W / \partial q_j \tag{1}$$

where W is a state function of q_j, internal state variables (ISVs) and temperature; additional parameters may also enter, such as moisture content. In thermodynamic terminology, W is the Helmholtz free energy. The basis for Equation (1) has been discussed by numerous authors (e.g. Coleman and Gurtin, 1967; Rice, 1971; Schapery, 1990).

The internal state variables are designated by $S_m (m = 1, 2, ... M)$. They serve to account for changes in the material's structure that cause *inelastic* behaviour such as cohesive and adhesive cracking, crazing and shear banding, as well as molecular-scale mechanisms such as large rotations of molecular segments in glassy polymers (Argon, 1973). For an arbitrary infinitesimal process with changes in q_j and S_m, and using the summation convention for repeated indices,

$$dW = \frac{\partial W}{\partial q_j} dq_j + \frac{\partial W}{\partial S_m} dS_m = Q_j dq_j - f_m dS_m \qquad (2)$$

where

$$f_m \equiv -\partial W / \partial S_m \qquad (3)$$

is a so-called *thermodynamic force*. (Unless stated otherwise in this paper, it is assumed for simplicity that other relevant parameters, such as temperature and moisture content, are constant.) Integrate Equation (2) from a reference state (at which $W = Q_j = 0$) to the current state along the *actual* process or path, to obtain the *total work* done on the body,

$$W_T \equiv \int Q_j dq_j = W + \int f_m dS_m \qquad (4)$$

As observed previously (Schapery, 1990), the second term cannot be negative because

$$\int f_m dS_m = \int_0^t TS' dt \geq 0 \qquad (5)$$

where $S' = f_m \dot{S}_m / T$ is the non-negative entropy production rate, T is absolute temperature, the overdot denotes a time derivative, and t is time. Equation (4) shows that the total work W_T is equal to the work of deformation W (without changes in the structure) plus the non-negative work of structural changes.

Equations for Limited Path-Independent Work

So-called evolution equations are used to predict changes in the ISVs. Those that are discussed next are motivated by the path-insensitivity of work observed during loading or unloading for composite laminates when strain rate-dependence is weak or nonexistent (Lamborn and Schapery, 1988, 1993).

Consider a specific set of processes (as defined by histories $q_j(t)$) and suppose that a specific set of the S_m changes. The total work is path-independent if and only if these S_m obey the ISV evolution equation

$$f_m = \partial W_s / \partial S_m \qquad (6)$$

where W_s is a state function of one or more S_m but is independent of q_j. The left side of Equation (6) may be interpreted as the available force for producing changes in S_m, while the right side is interpreted as the required force. (Observe also that Equation (6) contains the special case of crack growth when f_m is an energy release rate and $\partial W_s / \partial S_m$ is a critical energy release rate.) For the given set of processes, Equation (6) may not be satisfied for all M parameters; those S_m that do not satisfy Equation (6) are assumed constant and the equation for each of the associated m value is not used. The subscript p or q will be used in place of m to designate the parameters that change, which are taken to be P in number. From Equations (4) and (6) we find

$$W_T = W + W_s \qquad (7)$$

where W_s is taken as zero in the reference state. Thus, the total work is seen to be a function of state because W and W_s are functions of only the state variables. The second law of thermodynamics provides an inequality as a constraint on the changes in state,

$$\dot{W}_s = TS' \geq 0 \qquad (8)$$

Even if Equation (6) is satisfied for one or more S_p, this inequality may not allow them to change.

Instantaneous values of the S_p are such that they minimize the total work when the body passes through stable states; i.e., for $\delta S_p \delta S_p > 0$,

$$\partial W_T / \partial S_p = 0 \qquad (9)$$

$$\left(\partial^2 W_T / \partial S_p \partial S_q\right) \delta S_p \delta S_q > 0 \qquad (10)$$

In contrast to classical thermodynamic formulations (e.g. Fung, 1965), the total work W_T, *but not necessarily* W, is a minimum at each stable state.

It is observed that Equation (6) with $m = p$ represents p equations for finding the S_p as functions of q_j. Equation (10) guarantees that the S_p can be found (Lamborn and Schapery, 1993). Then, $W_T = W_T\left(q_j, S_p(q_j), S_k\right)$ where the S_k are the constant ISVs. From Equation (1),

$$Q_j = \partial W_T / \partial q_j \qquad (11)$$

showing that the body exhibits *hyperelastic* behaviour during the time any particular set of parameters S_p undergoes change. Because the total work is a potential in the q_j during inelastic processes, the incremental stiffness matrix $\partial Q_j / \partial q_i$ is symmetric. Conversely, given that this stiffness matrix is symmetric when one or more S_p change, then both Equations (6) and (11) follow (Schapery, 1990). The mathematical formulation and material or structural characterization used by Mast et al. (1992) are based on strains alone (i.e. on the hyperelastic representation of W_T) while in this paper the ISVs are explicitly used.

It should be emphasized that the work is path-independent only for those processes for which the same subset of S_m changes. Path-dependence exists when different subsets of S_m change on one path compared to another as, for example, paths for loading and unloading.

If forces act on sliding crack faces then they may have to be included in the set Q_j unless they are associated with frictionless contact; in the latter case, the effect of crack opening and closing can be taken into account through the form of the strain energy function. Coulomb friction, if significant, cannot be accounted for through a work potential, and therefore the stiffness matrix is not necessarily symmetric during processes involving crack face sliding. If, however, one can use a potential to characterize the relationship between crack-face forces and relative displacements between crack faces, Equation (11) can be extended to this case by including this potential (which may depend on additional ISVs) in W_T; such a simplification is applicable with surface free-energy effects that produce crack healing.

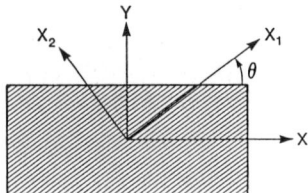

Fig. 2 Unidirectional composite and coordinate systems.

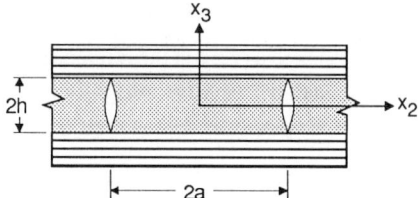

Fig. 3 Layer with transverse cracks. The cracked layer has fibres oriented in the x_1 direction.

3. Time-Independent Constitutive Equations for a Unidirectional Layer

The work potential theory reviewed in Section 2 is used here in the development of time-independent constitutive equations for a unidirectional layer. Coordinate notation for a layer is illustrated in Figure 2. Two ISVs will be used. One, designated by S, accounts for changes in the microstructure (i.e., on a scale that is much smaller than the layer thickness); the same resulting nonlinear, irreversible behaviour is assumed regardless of whether the layer is in a unidirectional or multidirectional laminate. The second ISV, designated by S_c, represents the effect of thermomechanically-induced transverse cracks. The planes of these cracks are, on the average, normal to the layer plane and parallel to the fibre direction, as illustrated in Figure 3.

These cracks typically span the full thickness of the unidirectional layer and, in brittle-resin carbon/epoxy composite tensile coupons, usually span the full specimen width. As such, it is common practice to use the reciprocal of average crack spacing, called *crack density* (or a related dimensionless parameter such as density times layer thickness) in representing their effect on laminate response (e.g. Daniel and Ishai, 1994 and Varna, 1992). However, the accumulated work/volume required to produce these cracks is a more convenient choice of an internal state variable for use in the work potential model. Furthermore, we assume that the effect of these cracks may be modelled through homogenized constitutive equations for a unidirectional layer, even though without adjacent layers having a different fibre angle the cracked layer would be in two or more pieces. According to the usual definition of a layer constitutive equation, the other layers in a multidirectional laminate have no effect on it; as discussed later, we find that this is an excellent approximation when S_c is work/volume, but not when it is crack density.

The ISVs, S and S_c, are assumed to be primarily responsible for the nonlinear layer behaviour during loading. The primary micro-mechanisms underlying S are not really known at this time; as noted earlier they may include matrix microcracks and shear banding, fibre-matrix debonds, and large rotations of molecular segments. However, both ISVs are associ-

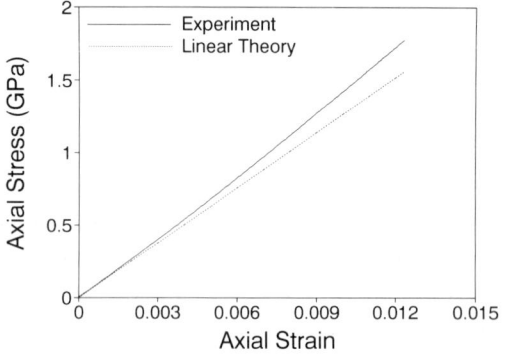

Fig. 4 Stress-strain behaviour of AS4/3502 unidirectional composite loaded in the fibre direction.

ated with reduced moduli during loading and unloading, compared to the initial moduli, and thus will be considered to provide measures of *damage*. For many fibre composites, the fibres themselves are another source of nonlinear behaviour. Assuming there is no significant fibre fracture, this nonlinearity produces an increase in modulus and decrease in Poisson's ratio for tensile loading in the fibre direction. An axial stress-strain curve is illustrated in Figure 4; this nonlinear behaviour is elastic, in that the axial stress and transverse strain curves are the same for loading and unloading (Schapery, 1989a).

Proposed Equations

Work potential-based constitutive equations which account for the effect of microdamage and elastic nonlinearity were developed and applied by Schapery (1989a). Using the $x_1 - x_2$ principal material coordinate system in Figure 2 and standard elastic property notation (e.g. Daniel and Ishai, 1994) they are,

$$\sigma_1 = Q_{11} f_1 \varepsilon_1 + Q_{12} f_{12} I_2$$
$$\sigma_2 = Q_{12} f_2 I_1 + Q_{22} f_2 \varepsilon_2 \qquad (12)$$
$$\tau_{12} = Q_{66} \gamma_{12}$$

where

$$Q_{11} = \frac{E_1}{1 - \nu_{12} \nu_{21}} \quad , \quad Q_{22} = \frac{E_2}{1 - \nu_{12} \nu_{21}}$$

$$Q_{12} = \nu_{12} Q_{22} \quad , \quad Q_{66} = G_{12} \qquad (13)$$

$$\nu_{21} = \frac{\nu_{12} E_2}{E_1}$$

Elastic nonlinearities appear through the following quantities,

$$I_1 \equiv \int_0^{\varepsilon_1} f_{12} d\varepsilon_1 \quad , \quad I_2 \equiv \int_0^{\varepsilon_2} f_2 d\varepsilon_2$$

$$f_{12} \equiv -\frac{1}{\nu_{12}} \frac{d\varepsilon_2}{d\varepsilon_1} \quad , \quad f_2 \equiv \frac{E_{90}(\varepsilon_2)}{E_{2_0}} \qquad (14)$$

$$f_1 \equiv \frac{E_0(\varepsilon_1)}{E_1}$$

The functions f_1 and f_2 are ratios of the secant modulus (E_0 or E_{90}) to the corresponding initial Young's modulus (E_1 or E_{2_0}) from uniaxial testing with loading in the fibre and transverse directions, respectively, without damage. These terms account for elastic nonlinearity in the fibre and transverse directions and represent nonlinearities that are observed both in loading and unloading. Elastic nonlinearity in the principal Poisson's ratio is modelled with f_{12}, a ratio of the tangent Poisson's ratio to the initial value ν_{12} from uniaxial tests with loading in only the fibre direction. These functions are drawn in Figure 5 for the AS4/3502 carbon/epoxy composite used in the experimental study; f_1 and f_{12} are functions of ε_1 and f_2 is a function of ε_2. Without elastic nonlinearity these functions are unity. The integral terms I_1 and I_2 are incorporated to satisfy the requirement that, for constant damage,

$$\frac{\partial \sigma_1}{\partial \varepsilon_2} = \frac{\partial \sigma_2}{\partial \varepsilon_1} \qquad (15)$$

which is a necessary and sufficient condition for the existence of a strain energy function. Since v_{12} is a constant (the initial Poisson's ratio), I_1 is the negative ratio of the transverse strain to the initial Poisson's ratio from a uniaxial test with loading in the fibre direction.

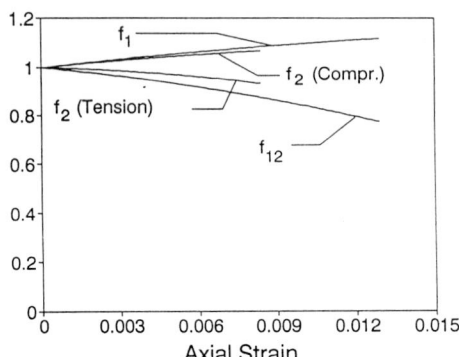

Fig. 5 Nonlinear elastic functions.

The effects of damage are accounted for through changes in the transverse modulus E_2 and principal shear modulus G_{12}. Although only the effect of S was characterized by Schapery (1989a), here we account for both S and S_c. Relative changes in these moduli due to damage are significantly larger than the nonlinear elastic behaviour implied by Figure 5. However, we shall not neglect the elastic nonlinearities because they have a measurable effect on the E_2 and G_{12} changes due to transverse cracking that are extracted from experimental data on behaviour of multidirectional laminates.

The Young's modulus in the fibre direction, E_1, and the Poisson's ratio, v_{12}, at zero axial fibre strain are assumed to be independent of damage, and thus are constants; elastic nonlinearities have been accounted for through the f − functions. Notice that v_{21} does change with damage growth through its dependence on E_2, as indicated by Equation (13).

The strains ε_1 and ε_2 are to be interpreted as the mechanical strains, i.e., the strains due to mechanical stress. Allowing for expansion (or contraction) strains due to temperature and/or moisture changes from the reference state, and using geometrically linear theory, then

$$\varepsilon_1 = \partial u_1 / \partial x_1 - \varepsilon_1^{ex}, \quad \varepsilon_2 = \partial u_2 / \partial x_2 - \varepsilon_2^{ex} \qquad (16)$$

where u_i are the in-plane displacements and ε_i^{ex} are the expansion strains. These expansion strains for the layer could conceivably be affected by damage, but we assume this is not the case here. It is implied by Equation (12) that the layer is mechanically orthotropic, with or without damage, and that the principal material directions are unaffected by the damage. The same simplicity is assumed for expansion strains, and therefore the shear strain is

$$\gamma_{12} = \partial u_1 / \partial x_2 + \partial u_2 / \partial x_1 \qquad (17)$$

The proposed Equation (12) may be identified with thermodynamic Equation (1) by taking $q_1 = \partial u_1 / \partial x_1$, $q_2 = \partial u_2 / \partial x_2$, $q_3 = \gamma_{12}$, $Q_1 = \sigma_1$, $Q_2 = \sigma_2$, $Q_3 = \tau_{12}$ and W as strain energy density. Integration yields the strain energy density (apart from an additive strain-independent function, which is taken as zero),

$$W = Q_{11}I_{11} + Q_{22}I_{22} + v_{12}Q_{22}I_1I_2 + \frac{G_{12}\gamma_{12}^2}{2} \qquad (18)$$

where

$$I_{11} \equiv \int_0^{\varepsilon_1} \varepsilon_1 f_1 d\varepsilon_1 \ , \ I_{22} \equiv \int_0^{\varepsilon_2} \varepsilon_2 f_2 d\varepsilon_2 \qquad (19)$$

In some applications one may want to account for geometric nonlinearities, such as large rotations and shearing strain. Assuming the form of Equation (12) does not change, this is easily done by replacing the displacement derivatives in Equation (16) and $\gamma_{12}/2$ in Equation (17) by Green's strains, and replacing the stresses in Equation (12) by Kirchhoff's stresses (Fung, 1965). Geometric nonlinearities were used by Schapery (1995) in the work potential theory for a study of local buckling due to compressive loading.

The combined effects of microdamage and transverse cracking on E_2 and G_{12} are approximated by assuming that the relative reduction in moduli due to transverse cracking is independent of the amount of microdamage. Equivalently, we suppose the effects of S and S_c appear through separate factors,

$$E_2 = E_{2_0} e_s(S) e_c(S_c) \qquad (20a)$$

$$G_{12} = G_{12_0} g_s(S) g_c(S_c) \qquad (20b)$$

where

E_{2_0}	= Transverse modulus at zero damage and zero strain.
$e_s(S)$	= Factor relating E_2 and the microdamage ISV, S.
$e_c(S_c)$	= Factor relating E_2 and the transverse cracking ISV, S_c.
G_{12_0}	= Shear modulus at zero damage and zero strain.
$g_s(S)$	= Factor relating G_{12} and the microdamage ISV, S.
$g_c(S_c)$	= Factor relating G_{12} and transverse crack ISV, S_c.

Our experimental results support Equation (20a); but for the laminates and loading used, they do not provide sufficient information on changes in G_{12} due to S_c to check Equation (20b).

Besides Equation (12), which relates stresses and strains, one needs two additional equations to be able to predict changes in the two damage ISVs. Equation (6) gives them, provided we know $W_s = W_s(S, S_c)$. It is assumed that the work of damage is additive, i.e. $W_s = W_1(S) + W_2(S_c)$, and that W_1 and W_2 are in one-to-one correspondence with their arguments. Then, without loss in generality, we may use for W_1 and W_2 the ISVs; i.e. take $W_1 = S$ and $W_2 = S_c$. Using evolution Equation (6) along with the definition of f_m, Equation (3), and Equation (18) for W, there results for $m = 1, 2$,

$$I_{11} \frac{\partial Q_{11}}{\partial S} + (I_{22} + v_{12} I_1 I_2) \frac{\partial Q_{22}}{\partial S} + \frac{\gamma_{12}^2}{2} \frac{\partial G_{12}}{\partial S} = -1 \qquad (21a)$$

$$I_{11} \frac{\partial Q_{11}}{\partial S_c} + (I_{22} + v_{12} I_1 I_2) \frac{\partial Q_{22}}{\partial S_c} + \frac{\gamma_{12}^2}{2} \frac{\partial G_{12}}{\partial S_c} = -1 \qquad (21b)$$

where we have used $S \equiv S_1$ and $S_c \equiv S_2$. (In practice it is usually possible to neglect the first term in these equations and use $Q_{22} = E_2$ since typically $v_{12} v_{21} \ll 1$). Further, Equation (7) for the total applied work reduces to

$$W_T = W + S + S_c \qquad (22)$$

which is useful in relating the ISVs to the strain in a characterization process, as discussed below. Notice that Equation (21) is equivalent to minimizing the work, Equation (22), for stable mechanical behaviour.

In Section 2 it was observed that the ISV evolution equations do not necessarily apply at all times. For example, referring to Equation (21), the strains may be too small for one or both equations to be satisfied; if an equation is not satisfied at the current values of the strains and ISVs, then the corresponding ISV does not change. Additionally, the entropy production rate cannot decrease, which implies from Equation (8) that $\dot{W}_s = \dot{S} + \dot{S}_c \geq 0$. Obviously, when only one ISV changes, it cannot decrease. However, considering the physical significance of S and S_c, we argue that neither can decrease, regardless of whether or not the other increases. Namely, the rate $\dot{S}_c$ reflects growth of transverse crack surface area; if there is no mechanism for healing of these cracks when $\dot{S} = 0$, it is not plausible that healing will occur when $\dot{S} > 0$ because the latter increase is associated with microscale processes distributed throughout the volume. Similarly, we argue that transverse cracking does not make microdamage healing possible.

As an illustration of these ideas, consider an initially undamaged unidirectional specimen which is subjected to a simple shear strain $\gamma_{12}(t)$ consisting of three segments: straining $(\dot{\gamma}_{12} > 0)$, unstraining $(\dot{\gamma}_{12} < 0)$ and restraining $(\dot{\gamma}_{12} > 0)$. Further, suppose that Equation (21a) is satisfied when the strain exceeds a certain value, and that it then predicts $\dot{S} > 0$. During the unstraining period Equation (21a) will initially predict $\dot{S} < 0$; but S cannot decrease, and thus it must remain at the largest value reached during the straining segment. If, during restraining, the earlier maximum strain is exceeded, then Equation (21a) would again be used as long as it predicts $\dot{S} > 0$. Similar considerations apply to Equation (21b) and S_c. (This type of analysis is, of course, analogous to what is used in plasticity theory when predicting plastic straining.) Equation (12) may or may not predict actual behaviour during unloading (when $\dot{S}$ and/or $\dot{S}_c$ vanish); if not, one or more additional ISVs may be introduced to characterize unloading behaviour (Schapery, 1989a).

Table 1. Laminates used in characterization and verification experiments

Unidirectional (12 plies):
$\theta = 0, 15, 30, 45, 90$

Angle-ply (12 plies):
$\pm\theta = 15, 30, 40, 45, 50, 60$

Multidirectional:

A $[\pm 45/60_2/\ast\pm 45/-60_2/\pm 45]_S$
B $[\pm 45/\pm 45/90/\ast\pm 45]_S$
C $[\pm 45/90_2/\ast\pm 45/90_2/\pm 45]_S$

An asterisk (*) indicates the location of an edge delamination due to loading.

Table 2. Elastic constants

Property	Value, GPa(Msi)
E_1	125.8 (18.25)
E_{20}	9.31 (1.35)
G_{120}	5.10 (0.74)
ν_{12}	0.329

Experimental Program

A carbon fibre-reinforced epoxy, Hercules' AS4/3502, with a 58% fibre volume fraction was used. The material was obtained in the form of unidirectional prepreg tape and cured in an air-cavity press using the manufacturer's recommended cure cycle; the cure temperature was 177C (350F). The laminates were produced in 30.5 x 30.5 cm (12 x 12 in.) plates and C-scan tested to screen for large flaws. Specimens were cut to approximately 1.27 cm (0.5 in.) wide, and were 19 cm (7.5 in.) long between the glass-epoxy end-tabs. They were stored in a desiccant at an ambient temperature of approximately 24C (75F) until tested. Each specimen was instrumented with two axial and two transverse strain gages, one set on the front and another on the back to average out any bending. Manufacturer's data were used to correct strain gage measurements for transverse gage sensitivity and nonlinearity. In all cases at least two specimens of each layup were tested, with very little specimen-to-specimen differences observed; that the scatter was very small was at least in part the result of not using specimens cut from plate edges (where the plates were thinner due to resin flow). The experimental stress-strain data from two replicas of each layup are shown later, starting with Figure 12. All results are for uniaxial tensile loading at a constant machine crosshead rate of approximately 1 mm/min. in the ambient environment.

Many different laminates were used in the experimental study in order to determine the material functions and to validate the results by using layups that were not used in the characterization process. Table 1 shows all layups that were used by Sicking (1992). Details on the experimental work and the characterization of material functions are covered elsewhere by Sicking (1992). Here, only a summary of the work is given.

Characterization of Thermoelastic Behaviour

Linear elastic material properties for the unidirectional composite were identified from uniaxial tests of $[0]_{12}$ and $[90]_{12}$ layups and a $[\pm 45]_{3S}$ angle-ply layup. Table 2 shows the linear elastic properties. The constants shown in this table represent the average of 2 or 3 test results and are first-order coefficients from a second- or third-order curve fit to the raw data. Recall that the

Table 3. Polynomial coefficients

Nonlinear elastic functions

Exponent	f_1	f_{12}	f_2 (tens.)	f_2 (compr.)
0	1	1	1	1
1	11.9	-14.6	-10.5	10.5
2	-214	-557	-156	-101

Damage dependent factors

Exponent	$g_s(S_r)$	$e_s(S_r)$ (tens.)	$e_s(S_r)$ (compr.)	$e_c(S_c)$
0	1.0	1.0	1.0	1.0
1	-0.03587	-0.0191	-0.0287	-0.00421
2	-0.01183	-0.00275	0.02232	2.68×10^{-6}
3	0.001401	-0.00025	-0.00342	0
4	-5.60×10^{-5}	2.54×10^{-5}	0.000146	0

nonlinear elastic functions are drawn in Figure 5. As described by Schapery (1989a) in an earlier experimental study of the same composite (but different lot), microdamage effects in the 90° coupons were separated from nonlinear elastic effects by observing the difference between loading and unloading curves. Approximately 75 percent of the variation in f_2 in tension was found to be retained when 90° unidirectional specimens were unloaded. The compressive portion of f_2 was determined from analysis of the 15° angle-ply specimens (in which $\varepsilon_2 < 0$), given the functions f_1 and f_{12}, and assuming that E_2 and G_{12} are constant; subsequent analysis of these specimens using the functions $E_2(S)$ and $G_{12}(S)$ confirmed this assumption since S was found to be practically zero. In the earlier work (Sicking, 1992) the compressive f_2 was assumed as unity, resulting in a questionable increase of E_2 with damage.

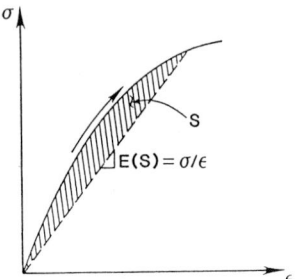

Fig. 6 Stress-strain curve for uniaxial loading in arbitrary (x) direction. The internal state variable S is equal to the shaded area. The dashed line is not necessarily the unloading stress-strain curve. For the case illustrated, the composite is linearly elastic when S is constant.

All nonlinear functions shown in Figure 5 are represented by second order polynomials in strain. Table 3 lists their coefficients as well as those for the damage-dependent factors which are discussed in the next subsection.

Thermal expansion strains, $\varepsilon_1^{ex} = \alpha_1 \Delta T$ and $\varepsilon_2^{ex} = \alpha_2 \Delta T$, were used to predict residual stresses and strains due to cooling from cure to ambient temperature ($\Delta T = -153 C$), in which $\alpha_1 = -0.54 \times 10^{-6} / C$ and $\alpha_2 = 22 \times 10^{-6} / C$ (Weaver, 1992). Small, competing effects due to chemical cure shrinkage (Fang et al. 1989) and moisture-induced expansion were neglected.

Characterization of Damage-Affected Behaviour

Microdamage: The damage-dependent factors in Equation (20) may be found in terms of the experimental specimen strains for each test. However, an objective is to express these functions directly in terms of S and S_c. Equation (22) provides the means for doing this when used with measured laminate responses. Consider, for example, the laminate stress-strain curve in Figure 6.

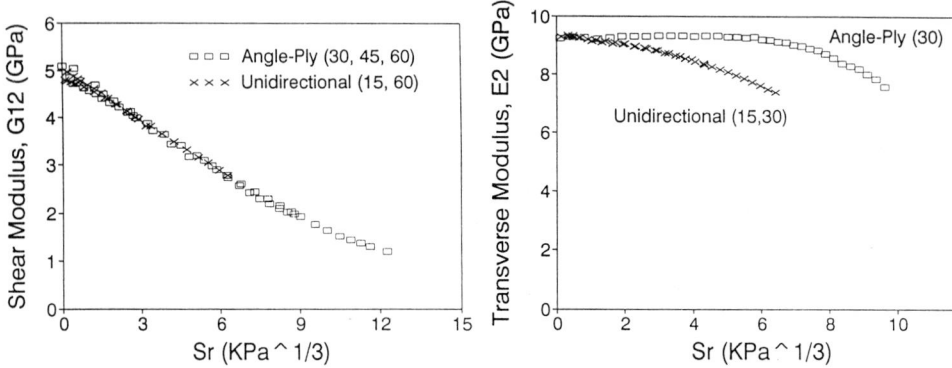

Fig. 7 Shear modulus versus reduced work of damage.

Fig. 8 Transverse modulus versus reduced work of damage.

It is illustrated for a unidirectional or multidirectional laminate without transverse cracking $(S_c = 0)$ and without elastic nonlinearity $(f_1 = f_2 = f_{12} = 1)$. For any given strain ε, the shaded area is $S = W_T - W$, and consequently S is easily found for each axial strain. With elastic nonlinearities, the lower curve is not straight, but Equation (18) provides the value of W. After one has obtained $e_s(S)$ and $g_s(S)$ from specimens without transverse cracks, other specimens with these cracks may be used to develop $e_c(S_c)$ and $g_c(S_c)$. At each axial strain Equation (22) gives $S_c = W_T - W - S$, where S may be predicted from Equation (21a).

It should be mentioned that E_2 and G_{12} were found to vary linearly in axial strain for very small strains. When $|\varepsilon| \ll 1$, then $W_T \cong W \approx \varepsilon^2$ and the difference without transverse cracking, $S = W_T - W$, behaves as ε^3. Thus, in order to express the moduli as polynomial functions, one should use $S^{1/3}$ as the expansion parameter. This was done for e_s and g_s in that we employed the *reduced* ISV,

$$S_r \equiv S^{1/3} \qquad (23)$$

in polynomial expansions. Use of this change-of-variable in Equation (21a) yields

$$I_{11}\frac{\partial Q_{11}}{\partial S_r} + (I_{22} + v_{12}I_1I_2)\frac{\partial Q_{22}}{\partial S_r} + \frac{\gamma_{12}^2}{2}\frac{\partial G_{12}}{\partial S_r} = -3S_r^2 \qquad (24)$$

Figures 7 and 8 present experimental G_{12} and E_2 data, respectively, from several different unidirectional and angle-ply specimens without transverse cracking. Division of the ordinate by the initial moduli G_{12_0} and E_{2_0}, respectively, provides g_s and e_s. Note the marked difference between E_2 data from 30° angle-ply specimens and from 15° and 30° unidirectional specimens. This behaviour is attributed to a difference in the sign of transverse stress σ_2 in these specimens; $\sigma_2 > 0$ for the unidirectional material and $\sigma_2 < 0$ for the angle-ply composite. The intuitively appealing conclusion from this finding is that the transverse compressive stress retards the transverse softening effect of microdamage, compared to the effect of a tensile stress. As a result, the work potential model has been formulated with two distinct polynomials for e_s, one for compression and another for tension. The best results have been obtained when 15° unidirectional and 30° angle-ply specimens were used to develop $e_s(S_r)$ for tensile and compressive behaviour, respectively. Response of these two types of specimens is the most sensitive to the material functions and it provides them over the largest S_r range, considering all laminates studied.

That the same functions, G_{12} and E_2, are found from different laminates (apart from the tension/compression difference in E_2) supports the assumption that these moduli are layer constitutive functions of one parameter, the work of microdamage. For comparison, the shear modulus is plotted against shear strain in Figure 9. This figure shows that G_{12} is *not* simply a function of shear strain.

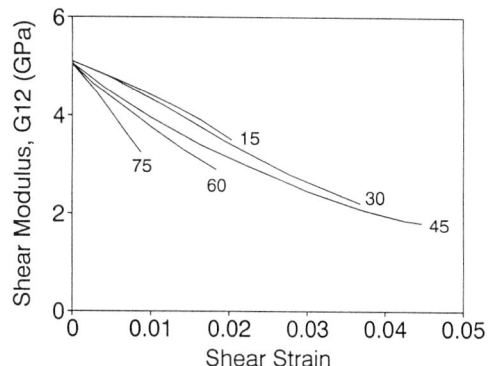

Fig. 9 Shear modulus versus shear strain from angle-ply laminates.

Transverse Cracking: The relationships of G_{12} and E_2 to the work of damage due to transverse cracking, S_c, can now be determined from specimens that exhibit transverse cracking. The basic approach is to use the work potential model to predict laminate behaviour up to transverse crack initiation. After transverse cracking begins, measured laminate stresses and strains are used along with predictions of the behaviour of layers without transverse cracks to determine the stresses, strains, and damage condition of layers with transverse cracks. Uniaxial tests give the values for laminate axial stress, σ_x, and normal strains, ε_x and ε_y, while other laminate stresses, σ_y and τ_{xy}, are zero. The Newton-Raphson method can be used to find the average stresses in each layer and microdamage state in uncracked plies. Details of this characterization process for obtaining e_c and g_c are given elsewhere (Sicking, 1992).

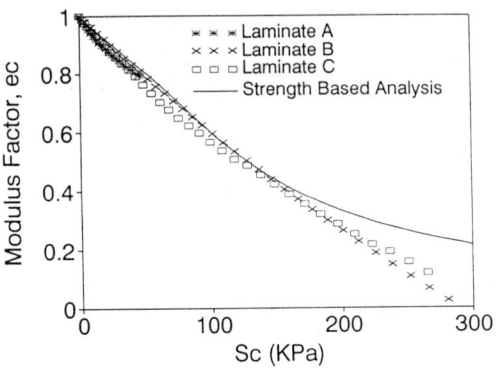

Fig. 10 Effect of transverse cracking on transverse modulus from theory and experiment. Transverse strength was used to predict cracking.

It was essential here to account here for residual thermal stresses due to cooling from the processing temperature of 177C (350F); results from different layups did not agree unless these stresses were taken into account. In contrast, the effect of thermal stresses on determination of e_s and g_s was very small, producing at most only a 1% and 2.5% change, respectively, in these functions.

Figure 10 shows the function e_c found from this process using the multidirectional laminates in Table 1, designated by A, B and C. Due to edge delamination laminate A failed at the lowest level of transverse cracking. Transverse cracks developed in the 60° layers in laminate A and the 90° layers in laminates B and C. The Tsai-Wu failure theory (Daniel and Ishai, 1994) was used to predict failure in the 45° layers; it indicated that failure (such as initiation of transverse cracking) does not occur until edge delamination initiates. Laminate A is thus the only one in which average shear stresses exist in the layers with transverse cracks, and hence it is the only one that can be used to find g_c; this function was found to be essentially unity over the S_c range that was induced. Figure 10 demonstrates that our material characterization provides a consistent relationship between reductions in E_2 and the work of transverse cracking. This consistency supports the assumption that e_c is a *ply* constitutive function, since essentially the same function is obtained for layers with both one and two plies and with different fibre angles.

As discussed in the introduction, there are many publications on the prediction of transverse cracking in *linear* elastic laminates. Here we use results from Hashin's (1985) analysis for comparison with our experimental results for e_c; inasmuch as the only source of nonlinearity in the analysis is transverse cracking, $e_c = E_2 / E_{2_0}$. He predicted the stiffness of crossply laminates as a function of crack spacing, but did not calculate the effective E_2 of the layers with transverse cracks. Nevertheless, this modulus can be extracted for application to the present study; the specific layups used for this extraction are discussed in the next subsection. Besides this step, another calculation is needed here because our characterization uses the work of damage as the measure of cracking. One approach is to assume that a transverse crack forms wherever the local strength is exceeded at a point midway between a pair of existing cracks, such as done by Lee and Daniel (1990) and Tsai and Daniel (1993). Another simple method is to assume the pre-existence of a dense spacing of initial transverse cracks throughout a unidirectional layer, in which the initial cracks span the full thickness $2h$ (cf. Figure 3), but are very short in the fibre x_1 direction. These cracks are assumed to not affect E_2 until the critical energy release rate is reached, after which crack growth occurs at a constant energy release rate, $G = G_c$, until the crack length equals the specimen width.. This second method was used by Nairn (1989) and Varna and Berglund (1991). We shall use both methods here to assess their accuracy when compared to the data in Figure 10.

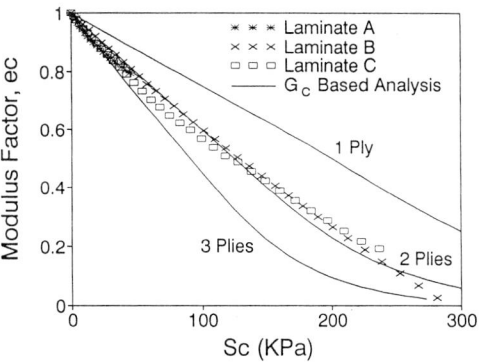

Fig. 11 Effect of transverse cracking on transverse modulus from theory and experiment. Critical energy release rate was used to predict cracking.

It should be added that the first method, which is based on strength, may be interpreted as a fracture mechanics approach in which the initial cracks are small compared to layer thickness; when these cracks become critical (as defined by a constant critical energy release rate $\hat{G}_c$, say) they propagate dynamically (because $G > \hat{G}_c$) until full-length transverse cracks are formed. Although a dynamic process is involved, path-independence of work is predicted under the assumption that the time elapsed for growth of each crack is negligible (Schapery, 1990).

The second method is the easiest to implement in the work potential theory. The instantaneous total work of cracking is equal to S_c times the volume of the cracked layer. This work is also equal to the constant critical energy release rate for transverse cracks, G_c, times the total instantaneous area of transverse cracks. These considerations lead immediately to

$$S_c = G_c / 2a \qquad (25)$$

where $2a$ is the instantaneous crack spacing (assumed the same for all cracks). Dimensional considerations imply the modulus is a function of crack spacing through only a dimensionless ratio (usually called the *normalized crack density*),

$$\beta \equiv h/a \qquad (26)$$

where $2h$ is the cracked layer thickness. From these two equations,

$$\beta = 2hS_c / G_c \qquad (27)$$

and thus E_2 is predicted to be a function of S_c and layer thickness. Figure 11 shows the predicted factor e_c for $G_c = 105$ N/m (0.6 lb/in.) and one, two and three plies in the cracked layer. There is clearly a significant effect of layer thickness, which is not exhibited by the experimental data; recall that the cracked layers in laminates A and C are two plies thick, while for laminate B the thickness is one ply. The value of $G_c = 105$ N/m was used as it provides the best agreement for the 2-ply case;

this value is close to our estimate of $G_c = 140$ N/m (0.8 lb/in.) for edge delamination, as discussed later. Equation (27) implies that if G_c is increased, the same predicted modulus moves to larger values of S_c.

The first method requires the stresses midway between cracks. Let us consider only the case in which $\tau_{12} = 0$. Then, denote the central tensile stress by σ_0, and write it in the form

$$\sigma_0 = f_0(\beta) E_{2_0} \varepsilon_2 \qquad (28)$$

The function f_0 will be found from Hashin (1985). In using Hashin's crossply analysis we introduce realistic simplifications for the cracked layer $v_{12}\varepsilon_1 \cong 0$ and $v_{12}v_{21} \cong 0$, and this $f_0(0) = 1$. (cf. Equations (12) and (13) for the linear elastic case). Referring to Figure 6, in which we use S_c in place of S, $\varepsilon = \varepsilon_2$, and $\sigma_0 = \sigma = E_2\varepsilon_2$, it follows that with $S = 0$,

$$S_c = W_T - W = \int_0^\varepsilon E_2(\beta')\varepsilon' d\varepsilon' - \frac{1}{2}E_2(\beta)\varepsilon^2 \qquad (29)$$

where $\beta' = \beta'(\varepsilon')$ in the integral. The connection between ε and β follows by setting $\sigma_0 = \sigma_f$ in Equation (28), where σ_f is the transverse tensile strength of a 90° laminate under a uniaxial tensile stress. We may then use Equation (28), with $\sigma_0 = \sigma_f$, to change the variable of integration to β' and bring out explicitly the effect of σ_f on S_c, for a given β. The result is

$$S_c = -\left(\frac{\sigma_f}{E_{2_0}}\right)^2 \left[\int_0^\beta E_2(\beta') f_0^{-3}(\beta') \frac{df_0}{d\beta'} d\beta' + E_2(\beta) f_0^{-2}(\beta)\right] \qquad (30)$$

Either Equation (29) or (30) relates β to S_c, and hence enables the transverse modulus to be predicted as a function of S_c. The resulting factor e_c is shown in Figure 10 using the transverse tensile strength of 69 MPa (10 ksi); this is the average value for our composite, as determined from $[90]_{12}$ coupons. Agreement is good for $0.4 < e_c \leq 1$. There is no predicted layer thickness effect, which is consistent with our data. Had an *in situ* strength of 66 MPa (9.6 ksi) been used in the prediction, then the theory would agree with the average of the experimental date for $e_c > 0.4$. In this study we found that the average *in situ* strength of 66 MPa is only slightly below the average strength of unidirectional specimens. However, this may not always be true (Varna, 1992), and depends on the accuracy of the transverse cracking model employed, the initial defects and possibly the material system.

The experimentally determined e_c falls below the theory at high crack densities, $h/a > 0.4 (e_c < 0.4)$, which is possibly due to local delamination between plies with transverse cracks and adjacent plies, crack branching (Varna, 1992) or cracks in the 45° layers. Inspection of failed specimens with such high crack densities indicated significant levels of this type of local delamination. When 90° layers are not on the surface of crossply laminates, this delamination is usually very small in monotonically loaded specimens (Jamison, 1986 and Varna, 1992). A large amount of local delamination may be the result of using angle-ply sublaminates here instead of 0° layers. In any event, the difference between theory and experiment for e_c at high crack density requires further investigation.

Upon comparing Figure 11 for only the 2-ply case with Figure 10, one might conclude that the G_c-based analysis is better than the strength-based analysis, and that the former analysis agrees with experimental data without the need to invoke an effect of local delamination or other mechanism. However, because experimental data are also shown for the one-ply case (laminate B), the absence of a thickness effect supports the strength-based criterion. This conclusion is consistent with experimental findings of Tsai and Daniel (1993).

Comparison of Analytical Models for Transverse Cracking

Although the effect of transverse cracks on the effective moduli of crossply laminates has been studied by many investigators, here we shall compare only Hashin's model (1985) with that of Lee and Daniel (1990); the former is based on the minimum complementary energy principle, while the latter is based on a shear lag analysis. (In earlier work, Sicking (1992) used Allen and Lee's (1991) prediction of E_2, but it was significantly above the experimental data; this is probably due to the use of an overly restricted axial displacement distribution, and resulting unrealistic axial stress distribution, in their potential energy-based analysis.) Hashin's analysis for crossply laminates was used to predict the solid line in Figure 10. We used $[0/90_n/0]$ laminates, with $n = 2$ and $n = 6$ to extract the factor e_c for the 90° layer; the two predictions cannot be distinguished graphically. Lee and Daniel's analysis was also used. For the S_c value at which Hashin's model predicts $e_c = 0.4$, their model yields $e_c \cong 0.4 \times 1.10$ and $e_c \cong 0.4 \times 1.18$ for $n = 2$ and $n = 6$ laminates, respectively. In contrast, when compared using normalized crack density h/a as the independent variable, instead of the strength-based S_c, there is a pronounced effect of n on the effective modulus factor e_c. Hashin's model predicts $e_c = 0.4$ and 0.4×1.2 for $n = 2$ and $n = 6$, respectively, for the same h/a value of approximately 0.4; Lee and Daniel's model predicts $e_c = 0.4 \times 1.4$ and $e_c = 0.4 \times 1.5$ for $n = 2$ and $n = 6$, respectively. It should be added that the predictions in Figure 11 are based on Hashin's model, with $n = 1$, 2 and 3. If E_2 as a function of h/a were independent of n, the spread in the curves would be noticeably larger than that in Figure 11.

In conclusion, we have found that both models predict that the effective modulus of the 90° layer is affected appreciably by the ratio of the number of 90° to 0° plies, when compared at the same normalized crack densities. On the other hand, when the strength-based S_c is used as the independent variable, the effect of n is much less; indeed it is negligible when Hashin's model is used, and e_c for the two models is closer than when normalized crack density is used. It would be instructive to make similar comparisons using Varna's (1992) more accurate model for effective modulus as a function of normalized crack density, and to consider a wider range of layups in order to determine the generality of our conclusions. Although this extended study has not been made, we are encouraged by the present experimental finding that the same e_c is obtained for the three different laminates, A, B and C and, as a function of the strength-based S_c, Hashin's model predicts essentially the same e_c for the symmetric crossply laminates with $n = 2$ and 6.

Additional Model Validation

A significant amount of model validation has already been done by demonstrating that different laminates provide essentially the same values of E_2 and G_{12}, as functions of S_r and S_c. Here, we show various predictions of stress-strain behaviour using constitutive Equation (12) and Equations (21b) and (24) to predict S_c and S_r, respectively. Nonlinear strain effects (Figure 5) as well as changes in E_2 and G_{12} due to microdamage (S_r or S) and transverse cracks S_c, using the product form in Equation (20), are accounted for. Figures 12-20 show experimental results for the

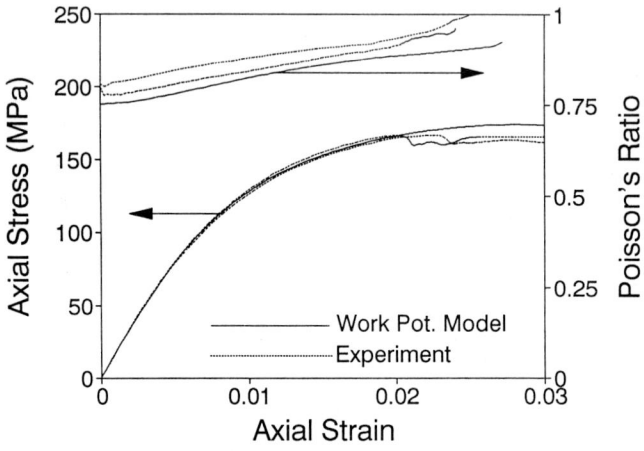

Fig. 12   Behaviour of [±45]₃ₛ laminate.

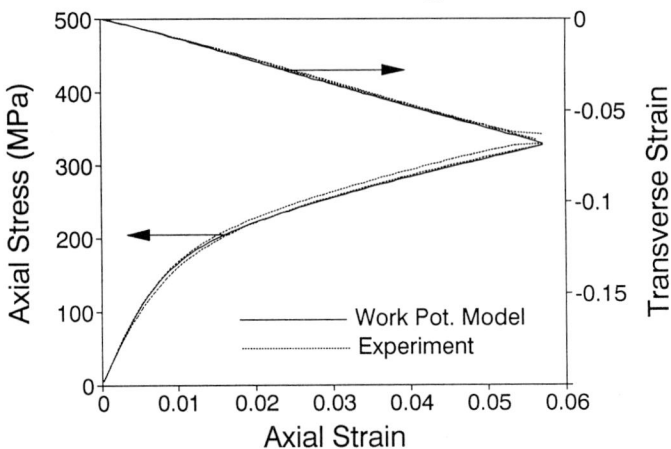

Fig. 13   Behaviour of [±40]₃ₛ laminate.

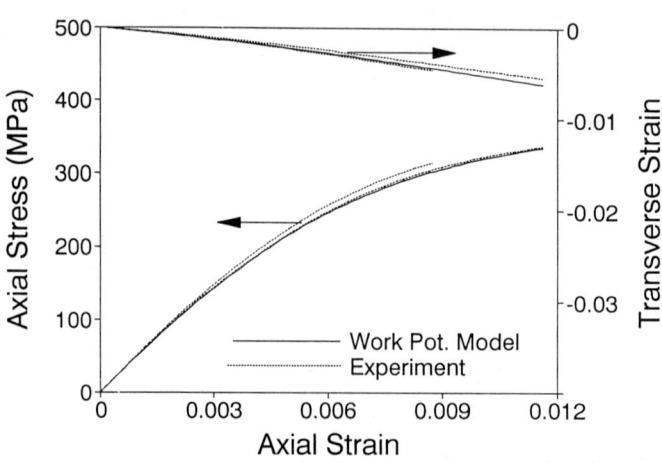

Fig. 14   Behaviour of [15]₁₂ laminate.

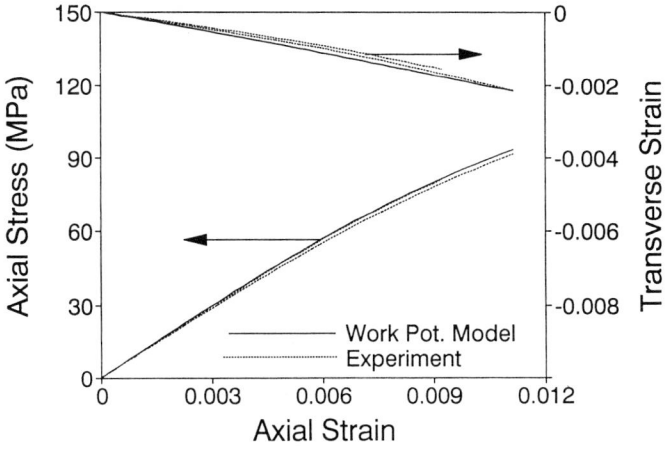

Fig. 15 Behaviour of $[60]_{12}$ laminate.

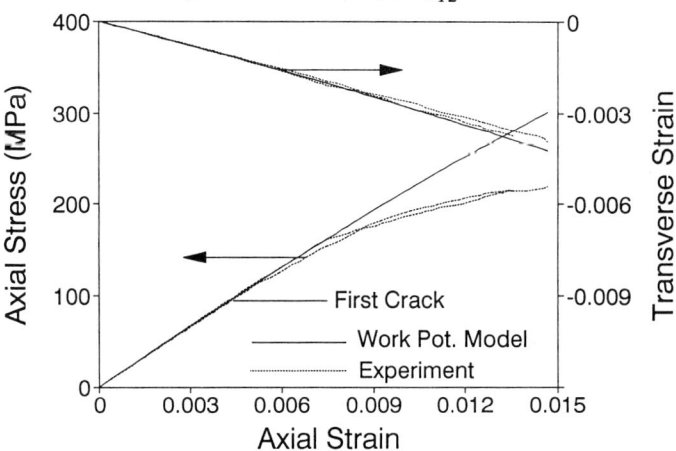

Fig. 16 Theoretical prediction of axial stress and transverse strain using $S_C = 0$ (without transverse cracks) for laminate C, compared to experimental results.

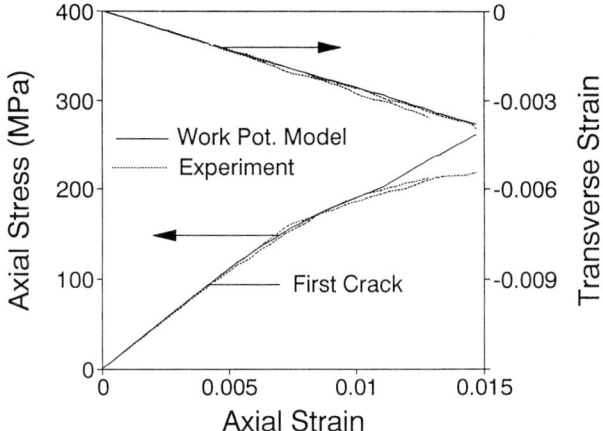

Fig. 17 Theoretical prediction of axial stress and transverse strain using $S_C \geq 0$ (with transverse cracks) for laminate C, compared to experimental results.

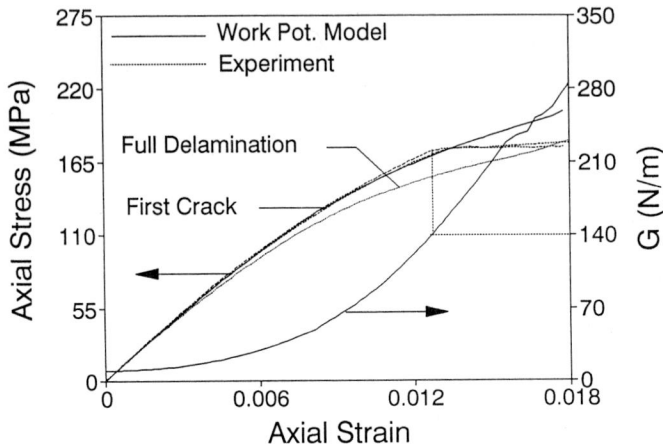

Fig. 18   Axial stress and delamination strain energy release rate for laminate A.

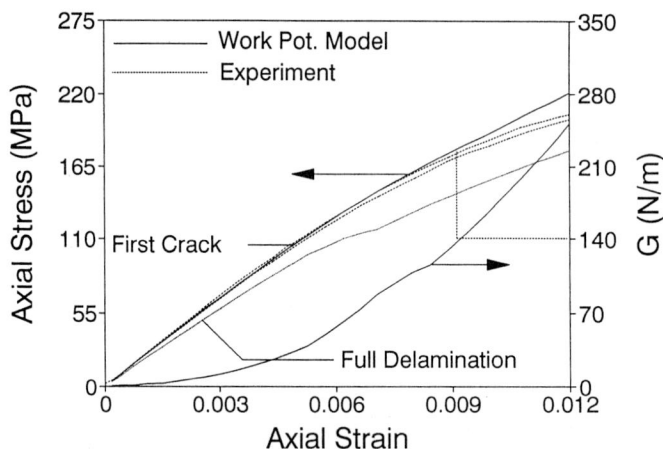

Fig. 19   Axial stress and delamination strain energy release rate for laminate B.

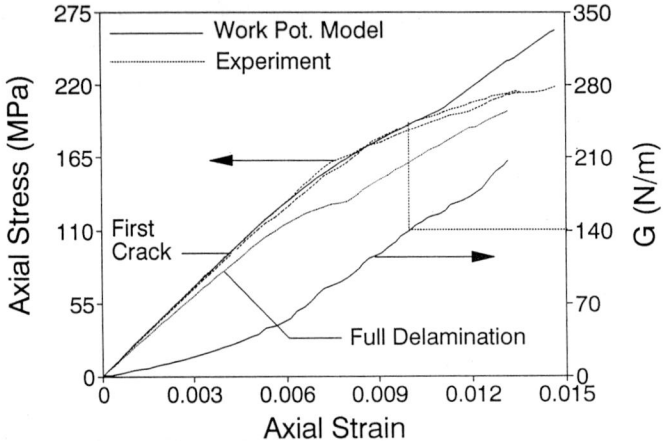

Fig. 20   Axial stress and delamination strain energy release rate for laminate C.

two tested replicas and show predictions for several laminates; axial strain was specified and then axial stress and transverse strain were predicted. Schapery (1989) and Sicking (1992) show predictions for several more laminates. In all cases agreement between theory and experiment is very good. Only a small amount of the experimental data were actually used in the characterization process. For example, rather than using the average of the G_{12} data in Figure 7 to obtain e_s, only the $[\pm 45]_{3s}$ laminate data was used. However, in view of the consistency of results discussed earlier, it would make little difference in the predictions if averages based on many laminates had been used instead.

It should be mentioned that the evolution equation for S_c, Equation (21b), is not satisfied during the early stages of loading since the magnitude of the left side is too small. Thus, there is a finite, critical applied strain at which S_c starts to increase. (Notice that the right side of Equation (24) increases continuously from an undamaged value of zero, and that the modulus derivatives are not zero; thus, softening due to S_r begins immediately with straining.) If there are some initial, large transverse cracks, one or more may grow according to the G_c-based model, but significant reductions in E_2 are not predicted until Equation (21b) is satisfied. This observation is consistent with the findings of Fang et al. (1989) where the G_c-based model was found to be in agreement with experimental results for only the initiation of transverse cracking. The reader is also referred to Varna (1992) for a theoretical and experimental investigation of the initiation of transverse cracking.

Figure 16 illustrates the effect of transverse cracking. The work potential theory in this figure does *not* account for the cracking, whereas it does in Figure 17. In Figures 17-20 the theory and experiment diverge at high stresses. This discrepancy is believed due to edge delamination which was observed to occur in all three multidirectional laminates, and is discussed in the next subsection.

Edge Delamination Analysis

Schapery (1989) implemented an adaptation of O'Brien's (1982) delamination analysis into the work potential model. The total work W_T for the laminate and sublaminates was used in place of the associated strain energies W in O'Brien's analysis. This formulation is applied here to determine strain energy release rates for delamination in the three multidirectional laminates shown in Table 1. Growing microdamage and transverse cracking are taken into account.

In the analysis edge delaminations are assumed to exist as depicted schematically in Figure 21. In contrast to the illustration, the actual extent of delamination was not fully symmetric, and the delamination was not uniform along the entire length of the specimens. However, there were at least patches of delamination that appeared to reflect the laminate's initial symmetry. For purposes of analysis we assume full symmetry, and a uniform depth a of delamination along each specimen's length. Additionally, following O'Brien, in order to easily predict the energy release rate G, it is assumed that the delamination depth is large compared to the sublaminate thicknesses so that G is independent of depth. These idealizations permit us to write the total work potential for a

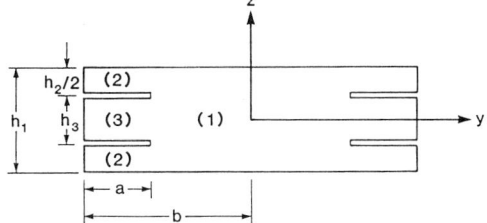

Fig. 21 End view of specimen with symmetrical edge delaminations. The loading is in the x-direction.

partially delaminated specimen in the form,

$$W_T = 2\{W_T^{(1)}(b-a)h_1 + [W_T^{(2)}h_2 + W_T^{(3)}h_3]a\}L + 4G_c aL \tag{31}$$

where $W_T^{(1)}$ is the work potential density of the central, undelaminated section and $W_T^{(2)}$ and $W_T^{(3)}$ are work potential densities of the separated sublaminates. The term $4G_c aL$ is the work of edge delamination, and represents the contribution of delamination to the work W_s introduced in Section 2. Also, b is the specimen half-width, a is the delamination depth, L is the length of delamination in the x-direction, h_i are the section thicknesses and G_c is the critical energy release rate for delamination. The delamination depth can be interpreted as an ISV. Thus, the delamination growth condition is $\partial W_T / \partial a = 0$ (cf. Equation (9)). This condition may be written in the equivalent form $G = G_c$, where G is the energy release rate for each delamination,

$$G = \frac{1}{2}[W_T^{(1)}h_1 - W_T^{(2)}h_2 - W_T^{(3)}h_3] \tag{32}$$

The one-half factor does not appear in the earlier work (Schapery, 1989a) because the thicknesses were one-half of those in Figure 21.

This expression was evaluated for all three laminates; It was found to be largest at the interfaces marked with an asterisk (*) in Table 1; these interfaces were used in evaluating G for each laminate. Delaminations were observed in test specimens at these interfaces as well as two plies from the surface in laminates A and C; these latter delaminations may be explained by the fact that G is only slightly lower than that for the locations marked in Table 1.

As indicated in Figs. 18-20, delamination is found to begin at a value of G of approximately 140 N/m (0.8 lb/in.). It is encouraging that the results are reasonably consistent, but it's surprising that it is less than the critical value for opening mode delamination reported for this material, from 161N/m (0.9 lb/in.) to 190 N/m (1.1 lb/in.), as measured from double cantilever beam (DCB) tests (Whitney et al., 1982, Bradley, 1989) on this composite; the edge-delamination is mixed-mode, and a value of G_c which is larger than from opening-mode tests should be expected (Bradley, 1989). However, the local delaminations associated with transverse cracking may have appreciably reduced the value of G_c from that reported for mixed-mode tests. Additionally, we predicted that transverse cracking has a negligible effect on G; but microdamage, through the parameter S, increases G at the onset of edge delamination by approximately 4 to 8 percent, depending on the laminate.

Also shown in Figs. 18-20 is the predicted stress-strain curve assuming that full delamination (i.e. $a = b$) exists from the beginning of loading. This curve provides an approximate lower bound to the stress-strain curve with edge delamination, and is seen to be close to the experimental curves in Figs. 18 and 20 at the highest strains.

4. Viscoelastic Deformation and Damage Growth Models

In this Section we extend the theory in Sections 2 and 3 to viscoelastic composites with growing damage. This work expands on an earlier theory (Schapery, 1981) by introducing explicit effects of high-modulus elastic fibres and transverse cracking. There seems to be few published studies on explicit analytical models for viscoelastic composites with damage. Zocher et al. (1994) analyse a linear viscoelastic laminate with a fixed number of transverse cracks. In a theoretical and experimental investigation, Park and Schapery (1994) characterize the behaviour of a particle-reinforced rubber with growing microcracks. Some models for nonlinear viscoelastic composites with growing damage are described by Schapery (1994); but here constitutive equations which are more detailed for fibre composites are developed.

Linear Viscoelastic Behaviour Without Damage

First, let us suppose there is no damage and the material is linearly viscoelastic and transversely isotropic. Looking ahead to be able to introduce certain simplifications and to account for out-of-plane stresses due to damage, we record the three-dimensional equations (referred to the principal material coordinates) in which strains are expressed as functions of stress history using compliances S_{ij} (Schapery, 1974),

$$\varepsilon_1 = S_{11}\sigma_1 + S_{12}(\sigma_2 + \sigma_3) \tag{33}$$

$$\varepsilon_2 = S_{12}\sigma_1 + \{S_{22}d\sigma_2\} + \{S_{23}d\sigma_3\} \tag{34}$$

$$\varepsilon_3 = S_{12}\sigma_1 + \{S_{23}d\sigma_2\} + \{S_{22}d\sigma_3\} \tag{35}$$

$$\gamma_{23} = \{S_{44}d\tau_{23}\}, \; \gamma_{13} = \{S_{66}d\tau_{13}\}, \; \gamma_{12} = \{S_{66}d\tau_{12}\} \tag{36}$$

$$S_{44} = 2(S_{22} - S_{23}) \tag{37}$$

The braces are abbreviated notation for the most general form of a linear hereditary integral,

$$\{fdg\} \equiv \int_{-\infty}^{t} f(t-t',t)\frac{\partial g}{\partial t'}dt' \tag{38}$$

When this integral is used with Equations (33)-(36), f is a compliance and g is a stress. If g is a constant applied at a time t_0, as in a creep test, Equation (38) reduces to

$$\{fdg\}/g = f(t-t_0,t) \tag{39}$$

for $t > t_0$, which is a creep compliance that depends on the time elapsed since the stress was applied, $t - t_0$, and the current time. In many cases the second argument can be omitted. The latter variable allows for chemical and physical aging (e.g. McKenna, 1994) as well as effects of transient temperature and moisture. For example, consider the familiar thermorheologically simple material,

$$\{fdg\} = \int_{-\infty}^{t} f(\xi - \xi')\frac{\partial g}{\partial t'}dt' \tag{40}$$

where

$$\xi - \xi' = \int_{t'}^{t} dt'' / a_T[T(t'')] \tag{41}$$

and a_T is the temperature shift factor. Observe that $\xi - \xi'$ is a function of t and t' or, equivalently, a function of $t - t'$ and t. If the temperature is constant then $\xi - \xi' = (t - t')/a_T$. As before, the strains are those due to stress. In order to account for thermal and moisture expansion effects, it is customary to simply use Equation (16) along with a similar equation for ε_3; the expansion strains are not usually sensitive to the history of temperature and moisture unless the temperature is close to the glass transition value (Ferry, 1980). Physical aging may be taken into account by introducing explicit time-dependence in a_T; i.e. use $a_T = a_T(T,t'')$ in Equation (41).

Referring to Equation (33), it is seen that the strain in the fibre direction is independent of stress history. This is an excellent approximation for the common case in which the fibres are elastic and have a much higher modulus than the matrix. Symmetry of the viscoelastic compliance matrix, $S_{ij} = S_{ji}$, then leads to the absence of a hereditary integral in the effect of σ_1 on ε_2 and ε_3.

A further simplification is possible by introducing the transverse creep Poisson's ratio v_{23},

$$v_{23} \equiv -\frac{\varepsilon_3}{\varepsilon_2} = -\frac{S_{23}}{S_{22}} \qquad (42)$$

where the strains are those due to a constant σ_2 alone. Assuming the composite deformation is dominated by that of the matrix, and accounting for the relative insensitivity of the matrix Poisson's ratio to time, as compared to time-dependence of the shear modulus (Schapery, 1974), then v_{23} is essentially constant. Thus, there are only two time-dependent compliances, S_{22} and S_{66}.

Next, let us invert Equations (33)-(36) so that a more convenient form for lamination theory is obtained. It is helpful first to develop an intermediate result by eliminating σ_1 in Equations (34) and (35) through the use of Equation (33), employing Equation (42) and then introducing the familiar relationship for axial modulus $E_1 = 1/S_{11}$ and principal Poisson's ratio $v_{12} = -S_{12}E_1$. We find that Equations (34) and (35) become, respectively,

$$\{S_{22}d(\sigma_2 - v_{23}\sigma_3)\} = \varepsilon_2 + v_{12}\varepsilon_1 + v_{12}^2(\sigma_2 + \sigma_3)/E_1 \qquad (43)$$

$$\{S_{22}d(\sigma_3 - v_{23}\sigma_2)\} = \varepsilon_3 + v_{12}\varepsilon_1 + v_{12}^2(\sigma_2 + \sigma_3)/E_1 \qquad (44)$$

The last right-side term is typically very small compared to the others in these equations, and thus will be neglected.

In order to invert these two equations we need the inverse form of Equation (38). Specifically, let

$$h = \{fdg\} \qquad (45)$$

Then,

$$g = \{f_I dh\} \equiv \int_{-\infty}^{t} f_I(t-t',t)\frac{\partial h}{\partial t'}dt' \qquad (46)$$

where it can be verified by direct substitution that f and its inverse f_I are related according to

$$\int_{t_0^-}^{t} f(t-t',t)\frac{\partial}{\partial t'}f_I(t'-t_0,t')dt' = H(t-t_0) \qquad (47)$$

where $H(t-t_0)$ is the unit step function $[H(t-t_0) = 0$ when $t < t_0$ and $H(t-t_0) = 1$ when $t > t_0]$. In general, f_I in Equation (47) is discontinuous when $t' = t_0$, and the lower limit t_0^- indicates that the integration is to include this discontinuity (through use of a Dirac delta function). When g is stress and h is strain, then f_I is a relaxation modulus. (For the common case in which $f = f(t - t')$ in Equations (38) and (45), then Equation (47) may be easily Laplace transformed if $t_0 \geq 0$, yielding the familiar result $p^2 \bar{f} \bar{f}_I = 1$, where p is the transform parameter and the overbar denotes a Laplace transform.) We now find Equations (33), (43) and (44) yield

$$\sigma_1 = E_1 \varepsilon_1 + v_{12} \{ E_2 d(\tilde{\varepsilon}_2 + \tilde{\varepsilon}_3) \} / (1 - v_{23}) \qquad (48)$$

$$\sigma_2 = \{ E_2 d(\tilde{\varepsilon}_2 + v_{23} \tilde{\varepsilon}_3) \} / (1 - v_{23}^2) \qquad (49)$$

$$\sigma_3 = \{ E_2 d(\tilde{\varepsilon}_3 + v_{23} \tilde{\varepsilon}_2) \} / (1 - v_{23}^2) \qquad (50)$$

where $E_2 = E_2(t - t', t)$ is the transverse relaxation modulus, which replaces the elastic transverse modulus E_2. Also,

$$\tilde{\varepsilon}_2 \equiv \varepsilon_2 + v_{12} \varepsilon_1 \quad , \quad \tilde{\varepsilon}_3 \equiv \varepsilon_3 + v_{12} \varepsilon_1 \qquad (51)$$

The shear relationships in Equation (36) become

$$\tau_{23} = \{ G_{23} d\gamma_{23} \} \quad , \quad \tau_{13} = \{ G_{12} d\gamma_{13} \} \quad , \quad \tau_{12} = \{ G_{12} d\gamma_{12} \} \qquad (52)$$

in terms of shear relaxation moduli G_{23} and G_{12}, where from Equation (37),

$$G_{23} = E_2 / 2(1 + v_{23}) \qquad (53)$$

Observe that Equations (49), (50) and the first of Equation (52) represent the constitutive equations for the $x_2 - x_3$ (isotropic) plane. Although ε_1 enters, its effect is like that of a thermal expansion or contraction strain.

The equations for plane stress, $\sigma_3 = \tau_{23} = \tau_{13} = 0$, become

$$\sigma_1 = E_1 \varepsilon_1 + v_{12} \{ E_2 d\varepsilon_2 \} \qquad (54)$$

$$\sigma_2 = \{ E_2 d(\varepsilon_2 + v_{12} \varepsilon_1) \} \qquad (55)$$

together with the last of Equation (52). For elastic behaviour these equations reduce to Equation (12) after setting $f_1 = f_2 = f_{12} = 1$ and neglecting the product $v_{12} v_{21}$ in Equation (13).

Correspondence Principle with Stationary or Growing Cracks

Equations (48)-(53) may be employed in the analysis of transverse cracking, and provide the basis for using a simple elastic-viscoelastic correspondence principle (CP) that is applicable with stationary or growing cracks. This CP does not use Laplace transforms, and is a special case of those developed by Schapery (1981, 1984) for linear and nonlinear viscoelastic media.

It is necessary to assume G_{12} is proportional to E_2 in order to use the CP. Then the only difference between a viscoelastic layer and an elastic layer lies in the use of a hereditary integral, with E_2 as the kernel function. In practice, G_{12} is only roughly proportional to E_2 (e.g. Lou and Schapery, 1971), but it is believed this does not cause significant error in the transverse crack problem.

By simply introducing a change of variables for displacements, the viscoelasticity problem is reduced to an elasticity problem, which has been called the *reference elasticity problem*. Start by defining so-called *pseudo-displacements* for $i = 1, 2$ and 3,

$$u_i^R = \frac{1}{E_R}\{E_2 du_i\} \tag{56}$$

and *pseudo expansion strains* for $i = 2$ and 3,

$$\Delta_i^R = \frac{1}{E_R}\{E_2 d\Delta_i\} \tag{57}$$

where E_R is an arbitrary constant with dimensions of modulus and

$$\Delta_i \equiv \varepsilon_i^{ex} - v_{12}\varepsilon_1 \tag{58}$$

Equations (48)-(50) and Equation (52) become,

$$\sigma_1 = E_1\varepsilon_1 + v_{12}E_R\left(\varepsilon_2^R + \varepsilon_3^R\right)/(1 - v_{23}) \tag{59}$$

$$\sigma_2 = \frac{E_R}{1 - v_{23}^2}\left(\varepsilon_2^R + v_{23}\varepsilon_3^R\right) \tag{60}$$

$$\sigma_3 = \frac{E_R}{1 - v_{23}^2}\left(\varepsilon_3^R + v_{23}\varepsilon_2^R\right) \tag{61}$$

$$\tau_{23} = \frac{E_R}{2(1 + v_{23})}\gamma_{23}^R \;,\;\; \tau_{13} = k_g E_R \gamma_{13}^R \;,\;\; \tau_{12} = k_g E_R \gamma_{12}^R \tag{62}$$

where $k_g \equiv G_{12}/E_2$. Also,

$$\varepsilon_2^R \equiv \frac{\partial u_2^R}{\partial x_2} - \Delta_2^R \;,\;\; \varepsilon_3^R \equiv \frac{\partial u_3^R}{\partial x_3} - \Delta_3^R$$

$$\gamma_{23}^R \equiv \frac{\partial u_2^R}{\partial x_3} + \frac{\partial u_3^R}{\partial x_2} \;,\;\; \gamma_{13}^R \equiv \frac{\partial u_3^R}{\partial x_1} + \frac{\partial u_1^R}{\partial x_3} \;,\;\; \gamma_{12}^R \equiv \frac{\partial u_2^R}{\partial x_1} + \frac{\partial u_1^R}{\partial x_2} \tag{63}$$

Equations (60)-(63) are identical to those for an elastic material with transverse modulus E_R, Poisson's' ratio v_{23} and with thermal expansion strains Δ_2^R and Δ_3^R. These constitutive equations are a special case of those used to establish the correspondence principle for quasi-static problems (with stationary or growing cracks) designated as CP-II by Schapery (1981, 1984). Equation (59) is needed as well in a full, three-dimensional formulation. However, ε_1, rather than pseudo strain, enters, which prevents an immediate application of CP-II. This difficulty will be dealt with here by assuming the fibres do not break and that the modulus E_1 is so high compared to E_2 and G_{12} that the strain ε_1 is essentially unaffected by the damage, given the laminate boundary displacements.

Let us now consider the specific problem of predicting the mechanical state of a unidirectional layer, in which the only damage is stationary or growing transverse cracks (cf. Figure 3). Although not necessary, it is conceptionally helpful to specify boundary conditions on the laminate entirely in terms of the displacements, say U_i. These are converted to pseudo displacements U_i^R through the hereditary integral, as in Equation (56). Of immediate interest is the development of the homogenized constitutive equations for a unidirectional layer with transverse cracks as well as a criterion for the growth of these cracks. If deformation of the layers adjacent to the cracked layer (in the neighbourhood of each crack edge) is significant, then it is assumed the constitutive equations of the adjacent layers can be reduced to elastic-like equations using the same transverse modulus as for the cracked layer (e.g., all layers may consist of the same composite material).

The solution to the reference elastic boundary value problem may be expressed as a sum of solutions to two problems: (1) that without cracks, given U_i^R (on the laminate boundaries) and $(\varepsilon_i^{ex})^R$ and (2) that with transverse cracks on which act specified tractions that are equal, but with opposite sign, to those existing at the same points in the uncracked layer; in this second problem the laminate boundary displacements and expansion strains vanish. Inasmuch as $\varepsilon_1 \cong 0$ in problem (2), it can be replaced by $\varepsilon_1^R \equiv \partial u_1^R / \partial x_1$ in Equation (59) without causing significant error. The combined tractions from (1) and (2) on the (outer) boundaries of the cracked layer are then found, resulting in the homogenized constitutive equations for the cracked layer.

Solution of problem (2) is obtained with CP-II by recognizing that all six Equations (59)-(62) are identical to elasticity equations. The viscoelastic stresses are equal to the elastic stresses, while the viscoelastic displacements are obtained by inverting Equation (56),

$$u_i = E_R\left\{S_{22}du_i^R\right\} \tag{64}$$

where S_{22}, as E_2 in Equations (56) and (57), is a material function for the uncracked layer. That these viscoelastic stress and displacement fields satisfy all of the governing field equations and boundary conditions (including equilibrium equations and traction conditions on the crack faces) can be readily verified by substitution. This result leads to the conclusion that the factors e_c and g_c for the viscoelastic layer, when expressed as functions of crack density, are identical to those for an elastic material.

It should be mentioned that quasi-static response has been assumed. However, transverse cracks often propagate dynamically prior to being arrested at ply interfaces or until the cracks extend the full width of the laminate. There may be initial, slow growth, followed by high-speed growth. It is usually reasonable to assume that the time for rapid propagation is negligibly small and that dynamic response, following crack arrest, is quickly damped out. If so, the effect of rapid propagation on layer response may be analysed by a time-wise step-function removal of tractions along the surfaces which form the fully developed crack faces. However, this type of quasi-static analysis is not explicitly needed in predicting overall or homogenized layer response since one may use elastic solutions, as described above.

If the fibres are viscoelastic and have a relatively high modulus, which is the case for aramid fibres (Wang et al., 1992), then the above analysis may be easily generalized to this case if v_{12} is constant. One merely replaces E_1 in Equation (59) by the appropriate hereditary integral.

Homogenized Constitutive Equations

A rigorous extension of Equations (12)-(14), together with Equation (20) for the moduli, for a homogenized viscoelastic layer is not possible using CP-II. However, we shall propose an extension that contains these time-independent equations and the CP-II-based linear viscoelastic equations with transverse cracking as special cases. Namely, replace ε_2 and γ_{12} by pseudo strains, $\hat{\varepsilon}_2 \equiv E_{2_0}^{-1}\left\{E_2^L d\varepsilon_2\right\}$ and $\hat{\gamma}_{12} \equiv G_{12_0}^{-1}\left\{G_{12}^L d\gamma_{12}\right\}$, respectively; the superscript "L" denotes undamaged linear viscoelastic relaxation moduli. The moduli E_{2_0} and G_{12_0} are suggested to be the same quantities that are in Equation (20), as determined from the same constant rate tests used to develop the f-functions shown in Equation (14). (Although relaxation moduli are needed in this formulation, relaxation tests are not needed to determine them; they may be calculated from functions generated by creep tests or other types of tests (Ferry, 1980).) The value of G_{12_0} has no effect on the shear stress in Equation (12) because it cancels out when the pseudo strain is used. However, the choice of E_{2_0} affects the function f_2, and thus influences the stress prediction. Of course, as long as f_2 is close to unity it makes little difference in how E_{2_0} is chosen; otherwise, the use of pseudo strain as an argument of f_2 and the choice of E_{2_0} must be evaluated experimentally. In general, further experimental studies for which viscoelastic effects are significant are needed to assess the proposed constitutive equations. Additionally, it is necessary to select ISVs and establish their evolution equations in order to complete the formulation of viscoelastic homogenized equations with damage; the next subsection is concerned with this aspect.

Rate-Dependent Damage Growth

Rate-dependence may be introduced while retaining the essential features of path-independent work. This was demonstrated by Schapery (1991b) for the case in which the S_m define changes in crack geometry and obey a power law in energy release rate with a large exponent. Whether the S_m define changes in crack geometry or in other structural changes, let us assume the evolution law for S_m is of the form (m not summed),

$$\dot{S}_m = cF_m f_m^{q_m} \tag{65}$$

where thermodynamic force $f_m > 0$, and $F_m = F_m(S_m)$; also, c is a positive, dimensionless function of time, possibly due to aging and transient temperature and moisture. The definition of f_m for viscoelastic material is discussed later in this subsection. Now, from Equation (65),

$$\hat{F}_m(S_m) \equiv \left(\int_0^{S_m} \frac{dS_m}{F_m}\right)^{1/q_m} = \left(\int_0^{\hat{t}} f_m^{q_m} d\hat{t}'\right)^{1/q_m} \tag{66a}$$

where $\hat{t}$ is a reduced time,

$$\hat{t} \equiv \int_0^t c(t')dt' \tag{66b}$$

As long as f_m does not decrease, the right side of Equation (66a) is approximately $k_m^{-1} f_m \hat{t}^{1/q_m}$ if $q_m \gg 1$, where k_m is a constant (and $k_m \to 1$ as $q_m \to \infty$). Thus,

$$f_m = k_m \hat{F}_m / \hat{t}^{1/q_m} \tag{67}$$

which replaces Equation (6). It is seen that if the contribution to W_s associated with those S_m which obey Equation (65) is taken as

$$\Delta W_s = \sum_m k_m \int_0^{} \hat{F}_m dS_m / \hat{t}^{1/q_m} \tag{68}$$

then essential results from the original time-independent theory apply. Whenever f_m decreases, the last of Equation (66a) is practically constant, and is given by its value at the time f_m started to decrease. Generally, for evolution laws of the form in Equation (65) (with the same or different exponents for each S_m), the behaviour is like that of a time- or rate-independent body except the work of structural change or damage is time-dependent. Namely, the work potential W_T is like that for an aging elastic material. This result is expected to apply when the evolution law for $\dot{S}_m$ is a strong function of f_m, even if it is not a power law. With cyclic loading a result similar to Equation (68) is found, except the number of cycles appears in place of time (Schapery, 1990).

If deformation viscoelasticity is not significant, then the thermodynamic force f_m is that used in Sections 2 and 3, viz. $-\partial W / \partial S_m$. On the other hand, if the constitutive equations are Equations (59)-(62), then additional considerations are needed to determine the appropriate representation of f_m. When $\dot{S}_m$ is crack speed, then according to viscoelastic fracture mechanics (Schapery, 1984) and the related damage theory (Schapery, 1991b), one uses $f_m = -\partial W^R / \partial S_m$, where W^R is a pseudo strain energy; if f_m is divided by the length of the crack edge, it becomes the viscoelastic J-integral. The potential W^R is the same as that for an elastic material, except pseudo strains (apart from ε_1), replace strains.

Representation of the effect of damage on moduli may be through the same functions e_s, e_c, g_s, and g_c as for the elastic composite, except S and S_c have to be reinterpreted. They can be found experimentally from *isochronal* curves, as in Figure 6, after the specimen's axial strain is replaced by axial pseudo strain if $\varepsilon_1 \approx 0$. When ε_1 is not negligible, a modified axial stress and/or axial pseudo strain must be used because ε_1 does not enter everywhere as a pseudo strain (e.g. see Equation (59)). To the authors' knowledge, this approach to modelling effects of damage growth in viscoelastic composites has not yet been applied to fibre composites. However, it has been successfully used for particle-reinforced rubber (Park and Schapery, 1994) using two ISVs; details are given therein on use of isochronal and constant rate data for characterizing damage-dependent moduli and relating the ISVs to axial pseudo strain.

5. Conclusions

A work potential has been shown to be capable of accurately modelling the time-independent mechanical behaviour of a carbon/epoxy composite during loading by accounting for microdamage and transverse cracking. The formulation employed in this paper expresses unidirectional layer stresses in terms of strains, because this is more convenient than the inverse form when used in laminate analysis. Although further validation of the procedure is warranted, it offers an approach for accurately modelling the behaviour of a wide range of fibre-reinforced plastic laminates.

It was shown that changes in homogenized unidirectional moduli due to microdamage and transverse cracks can be correlated using the associated work of damage as the characterizing parameter. The same dependence of moduli on work of damage was found from a variety of layups. It was further shown that a strength-based mechanics prediction of softening due to transverse cracks is in good agreement with a wide range of experimental results. Prediction of edge delamination in three different laminates using the overall laminate work potential was found to be consistent with the observed onset of delamination.

Finally, a method for extending the work potential model to allow for strain-based linear viscoelastic deformation effects and rate-dependent damage growth was outlined. Further experimental studies are needed in order to assess the proposed viscoelastic constitutive equations with damage growth. When nonlinear viscoelastic deformation effects are significant, constitutive equations for a unidirectional layer may be simpler using stresses, rather than strains, as the independent variables (Schapery, 1994).

Acknowledgement

The work of the first author was sponsored by the Office of Naval Research, Ship Structures & Systems, S&T Division, and the National Science Foundation through the Offshore Technology Research Center. This support is gratefully acknowledged.

References

Allen, D. H. and Lee, J.-W. (1991), Matrix cracking in laminated composites under monotonic and cyclic loadings, *Composites Engineering*, **1**, 319-334.

Argon, A. S. (1973), A theory for the low-temperature plastic deformation of glassy polymers, *Phil. Mag.*, **28**, 839-865.

Bradley, W. L. (1989), Relationship of matrix toughness to interlaminar fracture toughness, *Application of Fracture Mechanics to Composite Materials*, edited by K. Friedrich, Elsevier, New York, 159-187.

Budiansky, B. and Fleck, N. A. (1993), Compressive failure of fiber composites, *J. Mech. Phys. Solids*, **41**, 183-211.

Christensen, R. M. (1979), *Mechanics of Composite Materials*, Wiley-Interscience, New York.

Coleman, B. D. and Gurtin, M. E. (1967), Thermodynamics with internal state variables, *J. Chem. Phys.*, **47**, 597-613.

Daniel, I. M. and Ishai, O. (1994), *Engineering Mechanics of Composite Materials*, Oxford University Press, New York.

Fang, G. P., Schapery, R. A. and Weitsman, Y. (1989), Thermally-induced fracture in composites, *Engineering Fracture Mechanics*, **33**, 619-632.

Ferry, J. D. (1980), *Viscoelastic Properties of Polymers*, Wiley, New York.

Friedrich, K. (1989), *Application of Fracture Mechanics to Composite Materials*, Elsevier, New York.

Fung, Y. D. (1965), *Fundamentals of Solid Mechanics*, Prentice-Hall, Inc., Englewood Cliffs, NJ.

Hahn, H. T. and Tsai, S. W. (1973), Nonlinear elastic behaviour of unidirectional composite laminae, *J. Composite Materials*, **7**, 102-118.

Hashin, Z. (1985), Analysis of cracked laminates: a variational approach, *Mechanics of Materials*, **4**, 121-136.

Jamison, R. D. (1986), On the interrelationship between fiber fracture and ply cracking in graphite/epoxy laminates, *Composite Materials: Fatigue and Fracture*, ASTM STP 907, edited by H. T Hahn, 252-273.

Lamborn, M. J. and Schapery, R. A. (1988), An investigation of deformation path-independence of mechanical work in fiber-reinforced plastics, *Proceedings 4th Japan-U.S. Conference on Composite Materials*, Technomic Pub. Co., Lancaster, PA, 991-1000.

Lamborn, J. J. and Schapery, R. A. (1993), An investigation of the existence of a work potential for fiber-reinforced plastic, *J. Comp. Mat.* **27**(4), 352-382.

Lee, J.-W. and Daniel, I. M. (1990), Progressive transverse cracking of crossply composite laminates, *J. Composite Materials*, **24**, 1225-1243.

Lou, Y. C. and Schapery, R. A. (1971), Viscoelastic characterization of a nonlinear fiber-reinforced plastic, *J. Comp. Mat.*, **5**, 208-234.

Mast, P. W., Nash, G. E., Michopoulos, J., Thomas, R. W., Badaliance, R. and Wolock, I. (1992), Experimental determination of dissipated energy density as a measure of strain-induced damage in composites. Naval Research Laboratory report NRL/FR/6383-92-9369, Washington, D.C.

Masters, J. E. and Reifsnider, K. L. (1982), An investigation of cumulative damage development in quasi-isotropic graphite/epoxy laminates, *Damage in Composite Materials*, ASTM STP 775, edited by K. L. Reifsnider, 40-62.

McKenna, G. B. (1994), On the physics required for prediction of long term performance of polymers and their composites, *J. Research of the National Institute of Standards and Technology*, **99**, 169-189.

Moore, R. H. and Dillard, D. A. (1990), Time-dependent matrix cracking in cross-ply laminates, *Composites Science and Technology*, **39**, 1-12.

Nairn, J. A. (1989), The strain energy release rate of composite microcracking: a variational approach, *J. Composite Materials*, **23**, 1106-1129.

O'Brien, T. K. (1982), Characterization of delamination onset and growth in a composite laminate, ASTM STP 775, 140-167.

Park, S. and Schapery, R. A. (1994), A thermoviscoelastic constitutive equation for particulate composites with damage growth, Univ. of Texas Report No. SSM-94-3.

Rice, J. R. (1971), Inelastic constitutive relations for solids: an internal variable theory and its application to metal plasticity, *J. Mech. Phys. Solids*, 19, 433-455.

Schapery, R. A. (1974), Viscoelastic behaviour and analysis of composite materials, *Mechanics of Composite Materials*, **2**, edited by G. P. Sendeckyj, Academic Press, 85-168.

Schapery, R. A. (1981), On viscoelastic deformation and failure behaviour of composite materials with distributed flaws, *1981 Advances in Aerospace Structures and Materials*, **AD-01**, ASME, New York, 5-20.

Schapery, R. A. (1984), Correspondence principles and a generalized J integral for large deformation and fracture analysis of viscoelastic media, *Int. J. Fracture*, **25**, 195-223.

Schapery, R. A., Goetz, D. P. and Jordan, W. M. (1986), Delamination analysis of composites with distributed damage, *Proc. Int. Symp. Composite Materials and Structures,* edited by T. T. Loo and C. T. Sun, Technomic, Lancaster, PA, 543-548.

Schapery, R. A. (1987), Deformation and fracture characterization of inelastic composite materials using potentials, *Polymer Engineering and Science*, **27**, 63-76.

Schapery, R. A. (1989a), Mechanical characterization and analysis of inelastic composite laminates with growing damage, in *Mechanics of Composite Materials and Structures*, edited J. N. Reddy and J. L. Tapley, **AMD-100**, ASME, New York, 1-9.

Schapery, R. A. (1989b), A method for mechanical state characterization of inelastic composite laminates with damage, *Advances in Fracture Research,* **3**, ICF7, 2177-2189.

Schapery, R. A. (1990), A theory of mechanical behaviour of elastic media with growing damage and other changes in structure, *J. Mech. Phys. Solids* , **38**(2), 215-253.

Schapery, R. A. (1991a), Analysis of damage growth in particulate composites using a work potential, *Composites Engineering*, **1**, 167-182.

Schapery, R. A. (1991b), Simplifications in the behaviour of viscoelastic composites with growing damage, in *IUTAM Symposium on Inelastic Deformation of Composite Materials*, G. J. Dvorak (d), Springer-Verlag, New York, 193-14.

Schapery, R. A. (1994), Nonlinear viscoelastic constitutive equations for composites based on work potentials, *Mechanics USA 1994*, edited by A. S. Kobayashi, Appl. Mech. Rev., **47**, S269-275.

Schapery, R. A. (1995), Prediction of compressive strength and kink bands in composites using a work potential, *Int. J. Solids and Structures*, **32**, 739-765. Also Univ. Texas Report No. SSM-94-1.

Sicking, D. L. (1992), Mechanical characterization of nonlinear laminated composites with transverse crack growth, PhD Thesis, Texas A&M University, College Station, TX.

Sun, C. T. and Chen, J. L. (1989), A simple flow rule for characterizing nonlinear material behaviour of fiber composites, *J. Composite Materials*, **23**, 1009-1020.

Tsai, C. L. and Daniel, I. M. (1993), The behaviour of cracked crossply composite laminates under general in-plane loading, *Damage in Composite Materials*, edited by G. Z. Voyiadjis, Elsevier Science Publishers B. V., 51-66.

Tsai, C.-L., Daniel, I. M. and Lee, J.-W. (1990), Progressive matrix cracking of crossply composite laminates under biaxial loading, *Microcracking-Induced Damage in Composites*, edited by G. J. Dvorak and D. C. Lagoudas, **AMD-111**, ASME, New York.

Varna, J. (1992), Analysis of transverse cracking in composite cross-ply laminates, Doctoral Thesis, Dept. of Materials and Production Engineering, Lulea Univ. of Technology, Lulea, Sweden, Report 103D.

Varna, J. and Berglund, L. A. (1991), Multiple transverse cracking and stiffness reduction in cross-ply laminates, *J. Composites Technology and Research*, JCTRER, **13**, 97-106.

Wang, J. Z., Dillard, D. A. and Ward, T. C. (1992), Temperature and stress effects in the creep of aramid fibers under transient moisture conditions and discussion of the mechanisms, *J. Polymer Science: Part B: Polymer Physics*, **30**, 1391-1400.

Weaver (1992), private communication with representative of Hercules Aerospace Corporation.

Whitney, J. M., Browning, C. E. and Hoogsteden, W. (1982), "A double cantilever beam test for characterizing mode I delamination of composite materials, *J. Reinforced Plastics and Composites*, **1**, 297-313.

Zocher, M. A., Allen, D. H. and Groves, S. E. (1994), Analysis of the effects of matrix cracking in a viscoelastic composite at elevated temperature, *1994 Proc. 9th Conf. American Society for Composites*, Technomic Pub. Co., Lancaster, PA (in press).

HWANG Keh-Chih* and SUN Qing-Ping*

Thermoelastic Martensitic Transformation Induced Plasticity -
Micromechanical Modelling, Experiments and Simulations

Reference: Hwang, K.-C. and Sun, Q.-P. (1995), Thermoelastic Martensitic Transformation Induced Plasticity - Micromechanical Modelling, Experiments and Simulations. In: *Mechanical Behaviour of Materials* (ed. A. Bakker), Delft University Press, Delft, The Netherlands, pp. 77-100.

Abstact: Thermoelastic martensitic transformation induced plasticity (TMTRIP) is responsible for the behaviour of shape memory alloys and is also one of the basic mechanisms for crack tip shielding, i.e., toughening of some brittle structural ceramics. This review article attempts to organize and summarize the important research development in the area of constitutive relations and fracture of the martensitic phase transforming materials such as zirconia-containing ceramics and shape memory alloys. Prime concern is paid to the micromechanics constitutive description of the TMTRIP phenomena and its relations with transformation toughening. Recent advances in this area are emphasized, especially those in China.

1. Introduction

Phase transformations and other critical phenomena, as active research subjects in physics, constitute a field full of problems and unexpected discoveries. On the one hand, it is the subject of fundamental research in solid state physics, physical metallurgy and materials science; on the other hand, many phase transformable solids, such as shape memory alloys, steels and some kinds of ferroelectric and zirconia ceramics, have very important technical applications in modern high technology due to the fact that under external applied field these materials can exhibit some reversible mechanical and physical behaviours, among which, the *thermoelastic martensitic transformation induced plasticity* (TMTRIP) is most intensively studied from an applied mechanics point of view in recent years. The research effort on these materials in the last decade is mainly concentrated on the following two closely related basic aspects: (1) the constitutive description of transformation plasticity from a macroscopic or micro-macro combined point of view; (2) the influence of TMTRIP on the mechanical properties of materials such as the fracture, toughening and microstructural design. Therefore the research itself is principally a subject of interdisciplines.

The structure of this paper is as follows. First a brief description on martensitic phase transformations is provided in section 2. A review of some experimental and theoretical results on the constitutive relations of transformation plasticity, with prime concern paid to the micromechanics constitutive description of TMTRIP phenomena, is then given in section 3. As a result, some significant research contributions by other approaches, such as those of Tanaka's group in Japan and Fischer's group in Austria are not included. As an important aspect of constitutive research on TMTRIP, the research advances on transformation plastic flow localization (TPFL) phenomenon are described in section 4. Section 5 is devoted to the new research discoveries on toughening behaviour of these materials and the microstructural design of advanced materials.

---

* *Department of Engineering Mechanics, Tsinghua University, Beijing 100084, P.R. China*

2. Martensitic Phase Transformation and its Characteristics

2.1. Phase and phase transformations

A given assembly of atoms or molecules may be homogeneous or nonhomogeneous. The homogeneous parts of such an assembly, called phases, are characterized by thermodynamic properties such as volume, pressure, temperature, and energy. An isolated phase is stable only when its energy (or more generally, its free energy) is a minimum for the specified thermodynamic conditions. If the phase present is in a local minimum of the free energy instead of in a global minimum, and is separated from still lower minima (under the same thermodynamic condition) by energy barriers, the system is then said to be in a metastable state. If barriers do not exist, the state of the system becomes unstable and the system moves into a stable or equilibrium state, characterized by the lowest possible free energy. As the temperature, pressure, or any other variable like an applied stress, an electric field or a magnetic field acting on a system is varied, the free energy of the system changes correspondingly. Whenever such variations of free energy are associated with changes in structural details of the phase (atomic or electronic configurations), a *phase transformation* or *phase transition* is said to occur (see Rao and Rao, 1978). In this paper, our main concern is with the continuum mechanical aspects of *martensitic phase transitions* in solids. The methodology used here can be extended to the mechanics research of other kinds of phase transitions such as ferroelectric and ferroelastic transitions.

2.2. Martensitic phase transformation and its characteristics

Historically, martensite was the name given to the hard product obtained during the quenching of steels. The name was given in honour of the famous German metallurgist Martens (Wayman, 1964). A comprehensive definition of martensitic transformation is rather cumbersome. For example, a martensitic transformation can be defined as a structural change generated by atomic displacements and not achieved by diffusion, corresponding to a homogenous deformation which may be different in small adjacent regions and which gives rise to an invariant plane strain through which the parent and the product are related by a substitutional lattice correspondence, a habit plane (i.e., invariant plane) and a precise orientation relationship. However, it was found that the transformation to martensite was associated with certain characteristic structural features. Such features have now been observed in various non-ferrous metallic as well as non-metallic systems, such as *shape memory alloys* and *zirconia-containing ceramics*. These distinctive features include (Wayman, 1964; Christian, 1975): (1) Martensitic transformations are diffusionless in the sense that no thermally activated diffusion is required for the growth of the martensite and it is identical in chemical composition to the parent phase. During the transformation, the atomic positions in the parent phase change systematically and by distances less than the interatomic distances in the lattice. (2) The transformation occurs at a very rapid rate through a shearing of discrete volumes of the material. The two phases (martensite and the parent phase) are related by a deformational mechanism and show orientational relations. These features have led to the formulation of WLR (Wechsler, Lieberman and Read, 1953) phenomenological crystallographic theories of martensitic transformation. Here *"phenomenological"* means they provide a method of describing consistently *what* has happened, rather than *how* it actually happened. Even though the theories do not present a detailed picture of the actual process by which atoms move during the transformation, they are widely applied by many researchers and are used as the theoretical bases for the kinematical description of transformation plasticity. (3) Martensitic transformations usually are athermal, *i.e.*, they occur only when the temperature or the applied stress is changing. This behaviour occurs because the shape and volume changes associated with the change in crystal structure set up large strains which, due to the diffusionless nature of the transition, are not relieved by atomic migration. The resultant increase in the elastic strain energy opposes the progress of the transition, causing it to stop while incomplete; hence, a large driving force, which comes from undercooling or increase in applied external stress, is often required to induce further transformation. There are still some other characteristics in the thermodynamics and kinetics of martensitic transformation (see Rao and Rao,1978), but the shape and volume change mentioned above is of special importance. Among other things, it contributes to the *transformation plasticity* and to the *transformation toughening* in ceramics (Garvie *et al.*, 1975; McMeeking and Evans, 1982; Evans and Cannon, 1986; Sun and Hwang, 1994) and serves as the basis for their application as smart sensors and actuators. In the present paper the authors want to restrict themselves to thermoelastic martensitic transformations in metals and non-metals, since this kind of transformations are of specific relevance in many technical applications.

3. Constitutive Relations of Transformation Plasticity

Although the constitutive relations of materials have long been a very active research subject in the cross-field of solid mechanics and materials science, the study of the constitutive law for the phase transforming materials, however, is yet much less developed as compared with that of common metals. Shape memory alloys (SMA) and zirconia-containing ceramics are typical of metallic and nonmetallic thermoelastic martensites. The discovery of transformation toughening in ceramics, the wide application of shape memory alloys and the development of smart materials and structures have greatly promoted the constitutive research of TMTRIP in both theory and experiment. The major research advances in this active area in the last decade include: (1) the establishment of micromechanics constitutive theories and their applications to transformation toughening and transformation plastic flow localization (McMeeking and Evans, 1982; Budiansky *et al.*, 1983; Evans and Cannon, 1986; Sun, 1989; Sun, Hwang and Yu, 1991; Sun and Hwang, 1991; Sun and Hwang, 1990; 1993a; 1993b; 1994; Patoor *et al.*, 1988; Stam, 1994; Stam *et al.*, 1994; Yan, 1994); (2) the application of Landau-Devonshire theory of phase transition to the constitutive description of thermoelastic martensite (Falk, 1980; 1982; Falk *et al.*, 1990; Müller, 1989; Müller *et al.*, 1991; Song, 1994); (3) the application of finite thermoelastic theory to the displacive phase transition and the material instability analysis (see Abeyaratne, 1993). Of special significance is the current trend in research on the mechanics of phase transformations in solids to combine efforts from materials science, solid state physics and solid mechanics. This trend will greatly deepen and enrich our knowledge and understanding of TMTRIP and thus greatly push forward this interdiscipline.

The theoretical preliminaries for the establishment of the constitutive relations for transformation plasticity include: (1) a continuum mechanics and thermodynamics framework for constitutive relations governing transformation plasticity (Rice, 1971; 1975; Hwang, 1989). For example, a general internal-variable thermodynamics formalism may be employed to derive the constitutive law and set up the connection between macroscopic deformation and the underlying mechanisms operative on the microscale; (2) a crystallographic theory of martensitic phase transformations (see Wechsler *et al.*, 1953); (3) foundations of micromechanics and some basic approaches (such as the self-consistent approach) necessary for the micro-macro transition and for derivation of the free energy of the constitutive element; (4) Landau-Devonshire theory (see, e.g., Falk, 1980) of first order phase transition and its application to the martensitic transformations. In the micromechanics constitutive research of single and polycrystalline transforming solids, a constitutive element (representative material sample) is usually taken as the subject of study and the microstructure effects such as the crystallographic orientation, internal stress and energy dissipation *etc.* should be taken into account quantitatively. The derived constitutive law is usually required to be tenable for the general complex loading paths, so as to describe and to interpret the various phenomena by a unified interrelated model. With the higher levels of investigation and with new phenomena constantly being discovered (such as transformation localization), new knowledge and new methods are required. The recent advances in constitutive research are summarized as follows:

3.1. Micromechanics constitutive theory of transformation plasticity

The micromechanics constitutive theory of transformation plasticity aims at establishing an explicit quantitative relationship between the macroscopic behaviour of transforming solids and the underlying specific microstructure mechanisms and their evolution. Traditionally the constitutive behaviour of thermoelastic martensites can be classified under the following major headings : *The shape memory effect (Figure 1(c), (f)), the pseudoelasticity (Figure 1(a), (j)) and the ferroelasticity (Figure 1(d) (g))* (schematic illustrations of the behaviour are shown in Figure 1). Extensive microscopic studies have identified that the various forms of macroscopic behaviour under any thermomechanical loading are caused by one or combinations of the following crystallographic elementary processes at microstructural level (see the review papers by Delaey *et al.* (1974) for detailed description and physical interpretation): (1) thermoelastic parent to martensite ($p \rightarrow m$) forward transformation and its reverse transformation ($m \rightarrow p$); (2) reorientation from one martensite variant under applied stress to another kind of martensite variant ($m \rightarrow m$) (see Figures 2 and 3). Historically, micromechanics is closely related with phase transformations. The pioneering work of Eshelby (1956; 1957; 1970) on the elastic field of an ellipsoidal inclusion not only laid the foundation for micromechanics but also became the starting point in the constitutive research of transformation plasticity. The microstructural evolution processes typical of thermoelastic phase transforming solids provide ample research potential for the micromechanics. For example, one of the toughest

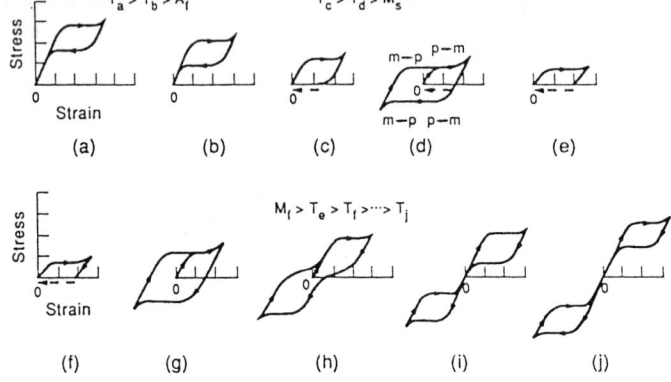

Fig. 1   Schematic illustration of the various stress-strain curves as a function of temperature for a typical shape memory alloy ($T_a > T_b > T_c > \ldots\ldots > T_i > T_j$).

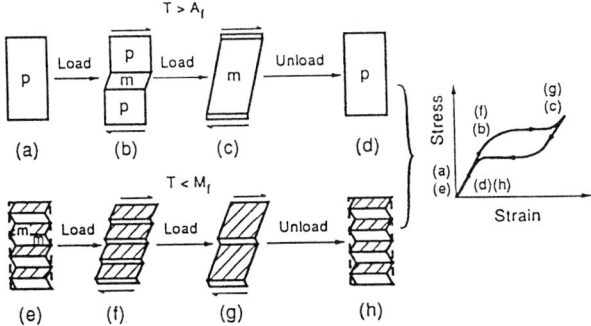

Fig. 2 Schematic illustrations of the mechanisms of pseudoelasticity due to transformation (a,b,c,d) and reorientation (e,f,g,h), respectively.

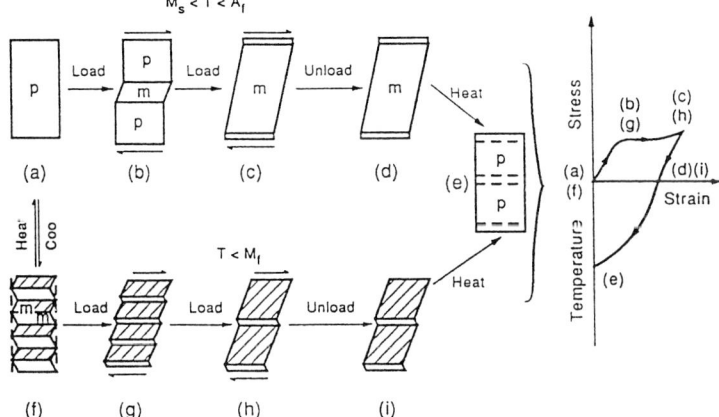

Fig. 3 Schematic illustrations of the mechanisms of shape memory effects (SME) by transformation (a,b,c,d,e) and reorientation (f,g,h,i,e), respectively.

problems is how to describe quantitatively the microstructure, its evolution and the interaction effects of stress fields due to individual transformed variants, which are very important in the derivation of the free energy of the material sample. The self-consistent approach is adopted and the research results can be summarized as follows:

3.1.1. Micromechanics constitutive model based on crystallographic theory

As a first fundamental step in establishing constitutive law of transformation plasticity, the micromechanics constitutive model of single crystals has been investigated by some researchers (Patoor et al., 1987; 1988; Yan, Sun and Hwang, 1994a). Starting from the lattice deformation of a martensite variant and by using the WLR crystallographic theory of transformation plasticity, Patoor, Eberhardt and Berveiller (1987; 1988) first derived the following kinematic equations for the *pure transformation plasticity* of thermoelastic martensite single crystals

$$\dot{E}_{ij}^{PT} = g \sum_n R_{ij}^n \dot{f}^n \tag{1}$$

where $\dot{E}_{ij}^{PT}$ is the macroscopic transformation plastic strain rate, g is the lattice constant of

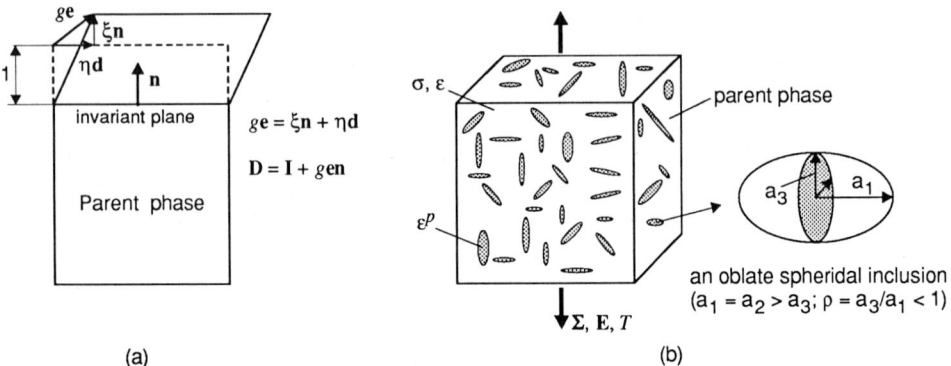

Fig. 4 Lattice deformation of parent phase (a) and the constitutive element of single crystal (b)

transformation, R_{ij}^n, f^n are the orientation tensor and volume fraction of the n th variant in a single crystal. They expressed the interaction energy among the variants as

$$E_{\text{int}} = Vg^2\left[\sum_n E^n f^n + \frac{1}{2}\sum_{n,m} H^{nm} f^n f^m\right] \quad (2)$$

where V is the volume of the constitutive element, E^n is the interaction energy between the single variant and the matrix, H^{nm} represents the interaction energy between different variants. However the explicit expressions for the H^{nm} have not been derived in the work of Patoor et al. (1988). The macroscopic constitutive relations can be finally expressed as (Σ_{kl}— macroscopic applied stress, B^m— material constants)

$$\dot{E}_{ij}^{PT} = \sum_n R_{ij}^n \sum_m H_{nm}^{-1}\left(R_{kl}^m \dot{\Sigma}_{kl} - \frac{B^m}{g}\dot{T}\right) \quad (3)$$

Based on this, the constitutive relation for polycrystals is derived by the self-consistent approach. With the assumption that only one variant of martensite is assumed to appear in a grain they calculated the stress strain curve during simple loading. Besides, a phenomenological yielding condition which is similar to the Drucker-Prager criterion is proposed to describe the macroscopic dilatant and shear effect during transformation, and it agrees well with the experimental data under proportional loading conditions.

Recently, Yan, Sun and Hwang (1994a) combined the crystallographic theory with the Mori-Tanaka self-consistent micromechanics approach (Figure 4), and derived the explicit expressions for the interaction energy and so obtained the analytic expression for the complementary free energy (negative Gibbs free energy) for unit volume of single crystals which is a function of applied macroscopic stress, temperature and the volume fractions ($f_1, f_2, ... f_N$) of the N kinds of variants ($N = 24$ for shape memory alloys)

$$\psi(\Sigma, T, f_1, \ldots, f_N)$$
$$= \frac{1}{2}\Sigma : \mathbf{M} : \Sigma + \Sigma : \sum_{s=1}^{N} \varepsilon_s^p f_s - (W + \Delta H) \sum_{s=1}^{N} f_s + \sum_{s=1}^{N} \sum_{t=1}^{N} f_s f_t \tilde{W}_{st} \qquad (4)$$

where Σ is the applied macroscopic stress, ε_s^p is the microstructral transformation strain corresponding to the formation of the sth kind of variants, T is the temperature, W is the elastic strain energy in an infinite medium due to a unit volume of oblate spheroidal inclusion with eigenstrain ε^p, $\tilde{W}_{st}$ represents the interactive energy between different kinds of inclusions. $\mathbf{M}$ is the elastic compliance tensor and ΔH is the chemical free energy change of unit volume from parent to martensite phase. The yielding surface for forward transformation of single crystal can be obtained (from analysis of energy dissipation rate) as the envelope of the following N hyperplanes in stress space

$$Y_s(\Sigma, T, f_1, \ldots, f_N) = \Sigma : \varepsilon_s^p - (W + \Delta H) + 2\sum_{t=1}^{N} \tilde{W}_{st} f_s - D_0 = 0, \quad s = 1, \ldots N \qquad (5a)$$

and the reverse transformation yielding surface is represented by the envelope of another set of N hyperplanes

$$Y_s(\Sigma, T, f_1, \ldots, f_N) = \Sigma : \varepsilon_s^p - (W + \Delta H) + 2\sum_{t=1}^{N} \tilde{W}_{st} f_s + D_0 = 0, \quad s = 1, \ldots, N \qquad (5b)$$

The reorientation process from any kth kind of variants to sth kind of variants is controlled by the following $N \times (N-1)$ hyperplanes

$$Y_{sk}(\Sigma, T, f_1, f_2, \ldots, f_N) = \Sigma : (\varepsilon_s^p - \varepsilon_k^p) + 2\sum_{t=1}^{N} f_t(\tilde{W}_{st} - \tilde{W}_{kt}) - D_0^r = 0, \quad s = 1, \ldots, N, \ k \neq s \quad (5c)$$

Thus totally the $N \times (N+1)$ hyperplanes dominate the constitutive response of single crystal under any complex loading conditions. The stress-strain relations obey normality rule and can be derived in the framework of internal variable theory. For example, the incremental form of the constitutive relation for forward transformation can be expressed as

$$\dot{\mathbf{E}} = \mathbf{M} : \dot{\Sigma} - \frac{g^2}{2} \sum_{s=1}^{N} \mathbf{R}_s \sum_{t=1}^{N} \tilde{W}_{st}^{-1} \left(\mathbf{R}_t : \dot{\Sigma} - \frac{k}{g} \dot{T} \right) \qquad (6)$$

where $k = \partial \Delta H / \partial T$ is material constant and the summation is only over the active variants. In the above work, the effect of the inclusion's shape on the elastic strain energy has also been analysed and the stress-strain relations under uniaxial tension is predicted quantitatively. The above constitutive relations can be also formulated in strain space (Sun, Yan and Hwang, 1994) and the condition for the uniqueness of the constitutive response is given. This model provides a theoretical foundation for the description of polycrystals.

3.1.2. Constitutive model of micromechanics mean field theory

The early constitutive models developed by McMeeking and Evans (1982), and by Budiansky et al. (1983) represent the pioneering work in this area. However, their models completely neglected

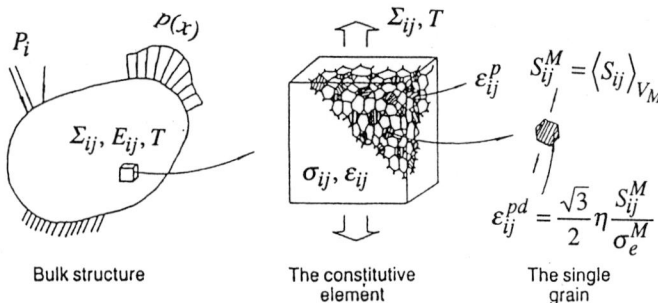

Fig. 5 *Microstructure of a polycrystalline constitutive element (left, middle) and its micromechanics idealization (right).*

the transformation induced shear strains associated with $t \to m$ transformations. This so-called "dilatant transformation" model has significantly enhanced the understanding of transformation toughening, but there is an unsatisfactory quantitative agreement with experiments (see, *e.g.* Evans and Cannon, 1986). Lambropoulos (1986) later revealed that the influence of transformation shear component on the shape of the transformation zone and toughening may be quite substantial, and his work has certainly triggered further research into the effect of transformation shear. Almost parallel to the theoretical work on transformation plasticity, significant experimental evidences for transformation shear were demonstrated by Chen and Reyes-Morel (1986), Reyes-Morel and Chen (1988), Reyes-Morel, Cherng and Chen (1988) and Sun, Huang, Yu and Hwang (1990), and their experiments all showed shear and dilatation effects of comparable magnitude. Based upon this, Sun, Hwang and Yu (1991) developed a new, micromechanics based continuum model of transformation plasticity for polycrystals with shear and dilatational effects. In the sense of the average of the microscopic field in a constitutive element, the self-consistent Mori-Tanaka theory (Mori and Tanaka, 1973; Mura, 1987; 1988) is employed to derive the expression for the interactions of the stress field of transformation (see Figure 5). The model of Sun *et al.* has been adopted to predict the constitutive behaviour of zirconia-containing ceramics and extended to shape memory alloys (Sun and Hwang, 1991; 1993a; 1993b; 1994) and is recently used to study the transformation toughening and transformation plastic flow localization (Sun and Hwang, 1992; Stam, 1994; Stam *et al.*,1994; Guo *et al.*, 1994). For thermoelastic martensitic polycrystals with nucleation control and stress-biased orientation (i.e., the assumption that deviatoric eigenstrain of grain is parallel to average deviatoric stress of matrix), the expression of the Helmholtz free energy of the constitutive element can be derived as

$$\phi\left(E_{ij},T,f,f^{re},<\varepsilon_{ij}^{pd}>_{V_{re}}\right) = \frac{1}{2}\left(E_{ij} - f<\varepsilon_{ij}^{p}>_{V_I}\right)M_{ijkl}^{-1}(T)\left(E_{kl} - f<\varepsilon_{kl}^{p}>_{V_I}\right)$$
$$-\frac{1}{4}B_1(T)\eta^2 f^{re} + \frac{1}{2}B_1(T)<\varepsilon_{ij}^{pd}>_{V_{re}}<\varepsilon_{ij}^{pd}>_{V_{re}}(f^{re})^2 \qquad (7)$$
$$-\frac{3}{2}B_2(T)(\varepsilon^{pv})^2(f - f^2) + 6\gamma_s f/d_0 + \Delta H(T)f$$

with $f<\varepsilon_{ij}^{p}>_{V_I} = f\varepsilon^{pv}\delta_{ij} + f^{re}<\varepsilon_{ij}^{pd}>_{V_{re}}$, where $\eta = \sqrt{3}(2\varepsilon_{ij}^{pd}\varepsilon_{ij}^{pd}/3)^{1/2}$ is the equivalent shear strain associated with the transformation, E_{ij} is macroscopic strain; T, temperature; f, volume fraction of martensites; f^{re}, volume fraction of reoriented martensite grains with long range shear; $<\varepsilon_{ij}^{pd}>_{V_{re}}$, deviatoric eigenstrain averaged over reoriented grains; V_I and V_{re}, the volumes occupied by the transformed martensite and the reoriented martensite, respectively; ε^{pv} the constant lattice volume dilatation. Other quantities in the above equation are either material constants or material functions of the independent variables given. Based upon the energy dissipation analysis (pre-knowledge of the evolution of the dissipated work as a function of the internal variables), the forward and reverse transformation conditions (both are deformation history dependent) of the constitutive element in stress space are derived as

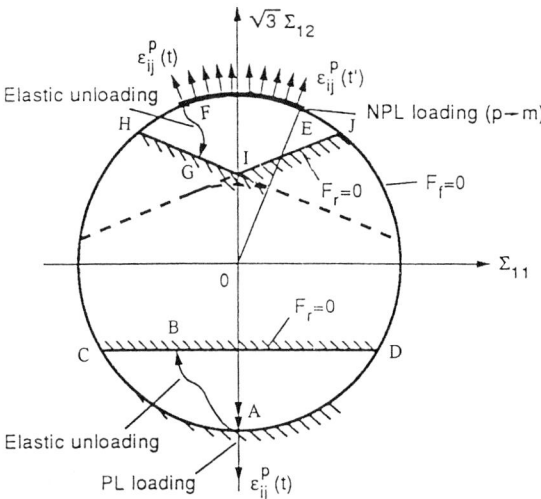

Fig. 6   Forward and reverse transformation surfaces under proportional loading (PL) and non-proportional loading (NPL) history.

$$F_f\left(\Sigma_{ij},T,f,f^{re},<\varepsilon_{ij}^{pd}>_{V_{re}}\right) = \frac{\eta}{\sqrt{3}} J\left(S_{ij} - f^{re}B_1(T)<\varepsilon_{ij}^{pd}>_{V_{re}}\right)$$
$$+ 3\varepsilon^{pv}\left[\Sigma_m - fB_2(T)\varepsilon^{pv}\right] - C_0(T,f) = 0; \quad (8)$$

$$F_r\left(\Sigma_{ij},T,f,f^{re},<\varepsilon_{ij}^{pd}>_{V_{re}},t'\right) = \left[S_{ij} - f^{re}B_1(T)<\varepsilon_{ij}^{pd}>_{V_{re}}\right]\varepsilon_{ij}^{pd}(t')$$
$$+ 3\varepsilon^{pv}\left[\Sigma_m - fB_2(T)\varepsilon^{pv}\right] - C_1(T,f) = 0, \quad (9)$$

where $J(\alpha_{ij}) = (3\alpha_{ij}\alpha_{ij}/2)^{1/2}$ and macroscopic stress is decomposed as $\Sigma_{ij} = S_{ij} + \Sigma_m\delta_{ij}$, t' is the past time parameter, $t' \in [0, t]$, t is current time. Equation (9) is the criterion of reverse transformation for the variant that was forward transformed at time t'. The forward transformation history is recorded by a memory function of $t'(t' \leq t)$. For the reorientation processes of the first and second kinds (see Sun and Hwang, 1993 a, b), the yield surfaces are derived, respectively, as

$$F_{ref}\left(\Sigma_{ij},T,f^{re},<\varepsilon_{ij}^{pd}>_{V_{re}}\right) = \frac{\eta}{\sqrt{3}} J\left(S_{ij} - f^{re}B_1(T)<\varepsilon_{ij}^{pd}>_{V_{re}}\right) - C_2(T,f) = 0 \quad (10)$$

$$\tilde{F}_{re}\left(\Sigma_{ij},T,f^{re},<\varepsilon_{ij}^{pd}>_{V_{re}},t'\right) = \frac{\eta}{\sqrt{3}} J\left(S_{ij} - f^{re}B_1(T)<\varepsilon_{ij}^{pd}>_{V_{re}}\right)$$
$$- \left[S_{ij} - f^{re}B_1(T)<\varepsilon_{ij}^{pd}>_{V_{re}}\right]\varepsilon_{ij}^{pd}(t') - D_0^{re2} = 0. \quad (11)$$

Detailed definitions for various quantities in the above equations are provided in the papers of Sun and Hwang (1993a; 1993b; 1994).

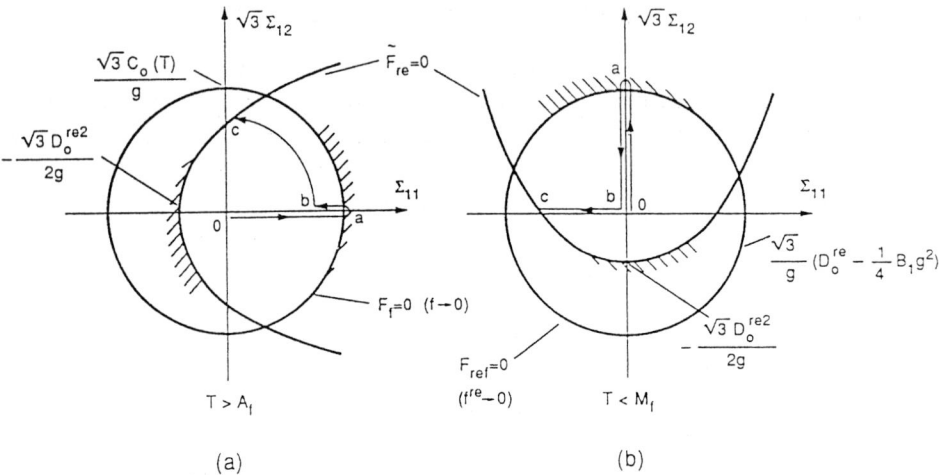

Fig. 7 Two-dimensional illustrations of the second kind of the reorientation process activated by NPL at high (a) and low (b) temperatures, respectively.

The yield surfaces of typical shape memory alloys under proportional and non-proportional loading histories in the two-dimensional stress space can be predicted by the present theory, as depicted in Figures 6 and 7. A corner structure of the yield surface for reverse transformation after a non-proportional transformation history is established in the figures. In Figure 6, CD represents the reverse transformation yield hyperplane after a proportional loading history OA, whereas the conical surface HIJ denotes the reverse transformation yield condition after a nonproportional loading history OEF. Figure 7(a) shows the second kind of reorientation yield surface (paraboloid $\tilde{F}_{re} = 0$) after a loading history oabc at temperature $T > A_f$, and Figure 7(b) shows the case at $T < M_f$. When used in the transformation toughening analysis of structural ceramics, the above constitutive model gives much better toughening prediction than BHL (Budiansky et al., 1983) model (see Sun, 1989: Sun, Huang, Yu and Hwang, 1991; Stam, 1994; Stam et al., 1994).

Another important application of this micromechanics constitutive theory to transformable polycrystals is the prediction of the materials behaviour related with transformation localization phenomena (Stam, 1994; Stam et al., 1994). For the macroscopic transformation localization of zirconia containing ceramics (such as Ce-TZP) induced by autocatalytic mechanism, this theory bears the following significance: (1) For pure dilatational and autocatalytic transformation of polycrystals with equal grain size, the bulk modulus during transformation is $-4\mu/3$, which is identical to the bifurcation condition (critical transformation of Budiansky et al.(1983)) with loss of ellipticity of the governing equation of dilatational plasticity. (2) For autocatalytic transformation with both shear and volumetric effects, the above theory can be used for the successful numerical simulation of the unstable growth of the narrow elongated transformation zone ahead of crack tip due to autocatalytic transformation, and henceforth provides an essential approach to estimate the undesirable reduction in crack tip shielding (see Guo et al., 1994). (3) According to the toughening theory based on controlling the spreading of localization (Marshall et al.1991; Marshall 1992), one can introduce the so-called dual-scale microstructure to control the direction of the localization and thus substantially increase the fracture toughness of ceramics. The constitutive model of Sun, Hwang and Yu (1991) provides a pertinent model for its quantitative realization (see sections 4 and 5 of this review paper).

3.1.3. Landau theory of phase transition and its applications to TMTRIP

Martensitic phase transitions are defined as diffusionless solid state structural phase transitions of first order with a deformation of the lattice such that a macroscopic strain results. Occasionally the first order condition is not included in the definition. Attempts have been made to describe martensitic phase transitions by means of the Landau-Devonshire theory by a number of researchers, among them the work of Falk (1980; 1982; 1990), Müller (1989) and Müller and Xu (1991) are most frequently cited. In 1980 Falk established a one-dimensional model on the constitutive relations of shape memory alloys, where the shear strain and shear stress were identified with the order parameter and the response, respectively. Some of the most important characteristics of the deformation features (such as lattice softening, pseudoelasticity, shape memory, ferroelasticity, *etc.*) are predicted qualitatively by this model. It can be concluded that the Landau-Devonshire theory or its extended version is one of the powerful universal thermodynamic approaches to the problems concerning martensitic transformations, because even such a one-dimensional model has demonstrated its power. Some researchers are encouraged by this and use this theory as one of the theoretical bases in establishing the constitutive model of transformation plasticity. As a phenomenological model, however, it has the following intrinsic shortcomings: (1) So far it is only successful in one dimension; recently attempts have been made to extend the theory to three dimensions (Falk and Konopka, 1990). Some difficulties still exist, one of which is the scale to be used in defining the relevant strain, for example in polycrystals the lattice deformation usually does not coincide with the macroscopic strain. (2) Though the theory is universal, far-reaching and deeply sighted, due to its phenomenological nature it is unable to take account of the microstructural effects such as the transformation induced internal stress, crystallographic orientation, shape of the martensite and interface friction *etc.* in the theoretical modelling. It seems that a possible way to overcome the above deficiency is to incorporate this thermodynamics theory, together with crystallography, microstructure and micromechanics into the continuum formulation of the constitutive description.

Recently a systematic research effort toward such a goal has been made (Song, 1994). In the work of Song (1994) the Landau-Devonshire theory of first order phase transition is incorporated into the micromechanics constitutive descriptions of polycrystals and combined with the crystallographic theory of martensite transformation. The forward and reverse transformations are treated as an instability in thermodynamic equilibrium and the variations of eigen strain and stress hysteresis with temperature are well described. Under the assumption of instantaneous transformation within a grain of polycrystal and the assumption that only one martensite variant is produced in a grain during transformation, the Gibbs free energy function of the constitutive element is derived as

$$G\left(\Sigma, T, f, <\boldsymbol{\varepsilon}^p>_{V_I}, <\boldsymbol{\varepsilon}^p:\boldsymbol{\varepsilon}^p>_{V_I}\right) = -\frac{1}{2}\Sigma:\mathbf{M}(T):\Sigma - \Sigma:f<\boldsymbol{\varepsilon}^p>_{V_I}$$

$$-\frac{1}{2}f\left[B_1<\boldsymbol{\varepsilon}^{pd}:\boldsymbol{\varepsilon}^{pd}>_{V_I} + 3B_2<(\varepsilon^{pv})^2>_{V_I}\right]$$

$$+\frac{1}{2}f^2 B_1 <\boldsymbol{\varepsilon}^{pd}>_{V_I}:<\boldsymbol{\varepsilon}^{pd}>_{V_I} \qquad (12)$$

$$+\frac{3}{2}<\varepsilon^{pv}>^2 f^2 + \Delta H^{p-m} f$$

where $\varepsilon^{pv} = \varepsilon^p_{ii}/3$, $\varepsilon^{pd}_{ij} = \varepsilon^p_{ij} - \varepsilon^{pv}\delta_{ij}$, ΔH^{p-m} is the difference in the Helmholtz free energy of the two stress free phases. According to the crystallographic theory of martensitic transformation and the Landau-Devonshire theory of first order phase transition, the following expressions can be redefined

$$\varepsilon^p = \frac{1}{2}\Delta g(\mathbf{en} + \mathbf{ne}) = \Delta g \mathbf{R}$$

$$\Delta g = g_m - g_p$$

$$\Delta H^{p-m} = H^m - H^p \quad (13)$$

$$= dg_m^6/6 - bg_m^4/4 + a(T-T_0)g_m^2/2$$
$$- [dg_p^6/6 - bg_p^4/4 + a(T-T_0)g_p^2/2]$$

where g is the order parameter defined as the displacement of atoms along vector $\mathbf{e}$, g_p and g_m are the order parameters of the parent and martensite phases, respectively (see Figure 8), d, b and a are material constants. After substituting Equation (13) into Equation (12) and replacing g_m by variable g, we can obtain $\partial G/\partial f|_{\Sigma,T}$, which can be considered as Gibbs free energy change during transformation of unit volume of parent phase to martensite. g_m and g_p should be calculated from the following equilibrium and critical conditions (denote $\partial G/\partial f|_{\Sigma,T}$ by $\Delta_f G$)

$$\frac{\partial(\Delta_f G)}{\partial g} = 0$$
$$\frac{\partial^2(\Delta_f G)}{\partial g^2} = 0 \quad (14)$$

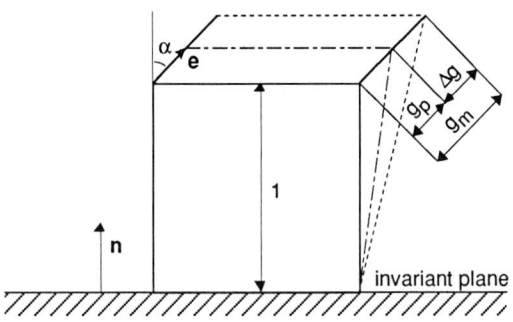

Fig. 8 Schematic illustrations of the lattice deformation during first order martensitic phase transition:
——— original stress-free parent phase lattice;
—·—·— the lattice of the parent phase just before the transformation;
- - - - - - the lattice after the transformation

$g_p = 0$ when the transformation is stress-free. The critical conditions (yield surface equation) for forward and reverse transformations can be finally derived respectively

$$\left(\Sigma - fB_1(T) < \varepsilon^{pd} >_{V_I}\right) : \mathbf{R} - fB_2(T) < \varepsilon^{pv} >_{V_I} \cos\alpha = dg_p^5 - bg_p^3 + a(T-T_0)g_p \quad (15)$$

$$\left(\Sigma - fB_1(T) < \varepsilon^{pd} >_{V_I}\right) : \mathbf{R} - fB_2(T) < \varepsilon^{pv} >_{V_I} \cos\alpha$$
$$+ \left[(\frac{1}{2} + \cos^2\alpha)B_1 + \frac{1}{3}B_2 \cos^2\alpha\right](g_{mr} - g_{pr}) = dg_{mr}^5 - bg_{mr}^3 + a(T-T_0)g_{mr} \quad (16)$$

where α is the angle between $\mathbf{e}$ and $\mathbf{n}$, the calculated g_{pr} and g_{mr} are respectively the values of g of the parent and martensite phases at the reverse transformation. The incremental constitutive relations can be obtained by normality rule. Theoretical predictions for the pseudoelastic stress-strain curve of shape memory alloys are shown in Figure 9.

This model overcomes some of the shortcomings of the previous models and is more universal. Some experimental phenomena such as the transformation curve (temperature-strain) during an autocatalytic transformation which the current existing models are unable to predict, can be now successfully explained by this new model (see Song, 1994).

3.1.4. Thermoelastic theory of phase transition and material instability

Thermoelasticity theory has been used to study certain general issues pertaining to solids that undergo reversible stress and temperature induced phase transformations. Various continuum-level issues related to reversible phase transformations in crystalline solids have been successfully studied using the theory of finite thermoelasticity (see Ericksen (1975) and James (1986)). For a thermoelastic material, the Helmholtz free-energy function ϕ depends on the deformation gradient tensor $\mathbf{F}$ and the temperature T: $\phi = \phi(\mathbf{F}, T)$. If the stress free material can exist in two phases, then the free energy function must have two separate energy-wells, each well corresponding to one phase. At $T = T_c$, the two minima have the same value; for $T > T_c$ the austenite minimum is smaller, while for $T < T_c$, the martensite minimum is smaller. A complete constitutive description of the material consists of three ingredients: a Helmholtz free energy function which describes the response of each individual phase, a nucleation criterion which signals the conditions under which the transition from one phase to another commences, and a kinetic law which characterizes the rate at which this transition progresses.

Much recent activity in continuum mechanics studies on thermoelastic phase transitions has been focused on two basic issues: the first one concerns energy minimizing deformations corresponding to the stable configurations of a body; the second is related to the non-equilibrium evolution of a body towards such stable configurations through intermediate states of metastability. The above two issues have been comprehensively summarized by Abeyaratne(1992). A continuum constitutive model for thermoelastic solids has been recently constructed by Abeyaratne et al.(1993). The free energy is associated with a three-well potential energy function; the kinetic relation is based on thermal activation theory; nucleation is assumed to occur at a critical value of the appropriate energy barrier. The predictions of the model in various quasi-static thermomechanical loadings are examined and compared with experimental observations.

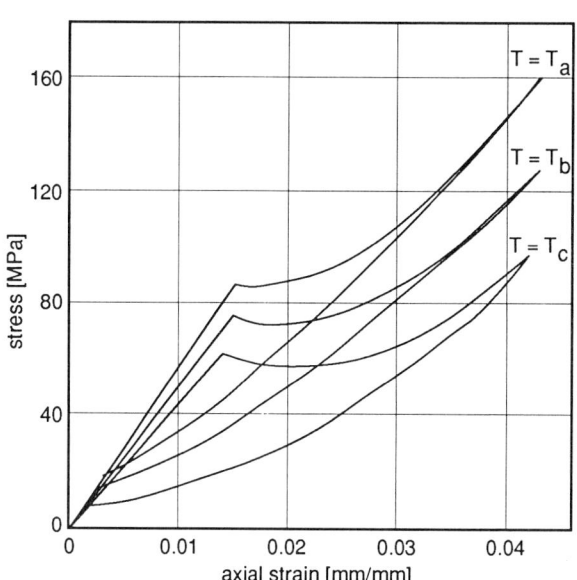

Fig. 9 Pseudoelastic stress-strain curves of a typical SMA predicted by combining micromechanics with the Landau-Devonshire theory $(T_a > T_b > T_c)$; after Song (1994)

4. Autocatalytic Transformation and the Induced Plastic Localization

The discovery of transformation plastic flow localization phenomena in ceramics and shape memory alloys stimulated a lot of research in the last few years. The transformation localization in ceramics is considered to be induced by an autocatalysis effect in the process of transformation. From the point of view of materials science, the autocatalytic transformation is initiated by a multiple-nucleation event (Chen and Reyes-Morel, 1986; Reyes-Morel and Chen, 1988; Reyes-Morel, Cherng and Chen, 1988) that, once realized, can stimulate further transformation to propagate rapidly over an extended region. The transformation in Ce-TZP ceramics is thermoelastic in nature, so the transformation induced internal stress and the corresponding stored elastic strain energy will have very important influence on the thermodynamics and kinetics of transformation (Delaey et al., 1974; Reyes-Morel and Chen, 1988; Sun and Hwang, 1994). From the micromechanical consideration, the average internal stress in the matrix caused by transformation is conducive to the further transformation of the remaining grains of the parent phase in the matrix. This kind of internal stress effect is most obvious in the case of equal grain size distribution (so the equal potency of nucleating sites in each grain). It is clear that this chain reaction, once initiated, is expected to require somewhat lower driving force to sustain, hence resulting in an externally applied stress decrease (the macroscopic load drop as observed in experiments).

4.1. Constitutive Description of Autocatalytic Transformation

The micromechanics constitutive model accounting for such important effect with the typical softening stress-strain response has been established by Sun, Hwang and Yu (1991). This model is used as the starting point for the localization analysis and the numerical simulation of the resulting crack-tip morphology. The macroscopic incremental stress-strain relation can be expressed as

$$\dot{\Sigma}_{ij} = L_{ijkl} \dot{E}_{kl} \qquad (17)$$

where the tangent modulus tensor can be derived to be

$$L_{ijkl} = L^o_{ijkl} - \frac{(L^o_{ijmn} T_{mn})(T_{pq} L^o_{pqkl})}{h + T_{ab} L^o_{abcd} T_{cd}} \qquad (18)$$

in which L^o_{ijkl} is the elastic stiffness tensor, $T_{ij} = \varepsilon^{pv} \delta_{ij} + A s^M_{ij} / \sigma^M_e$, $A = \sqrt{3\varepsilon^{pd}_{ij} \varepsilon^{pd}_{ij}/2} = \sqrt{3}\eta/2$, $h = 2B_1 A^2/3 + 3B_2 (\varepsilon^{pv})^2 + \alpha B_0 (\varepsilon^{pv})^2$. Detailed meaning of the parameters in the above expressions are given in references (Sun, Hwang and Yu, 1991; Stam, 1994).

4.2. Localization analysis

The term localization refers to the situations in which the deformations concentrate into a band as an outcome of the constitutive behaviour of the material. The orientation of the band is characteristic of the material, rather than a consequence of the boundary conditions. According to Rice (1976), localization of plastic flow is a special type of material instability, which is associated with loss of ellipticity of the governing differential equations. For the constitutive law described in Equations (17) and (18), the localization condition can be written as

$$A_{jk}(\mathbf{n}) m_k = (n_i L_{ijkl} n_l) m_k = 0, \quad j = 1, 2, 3 \qquad (19)$$

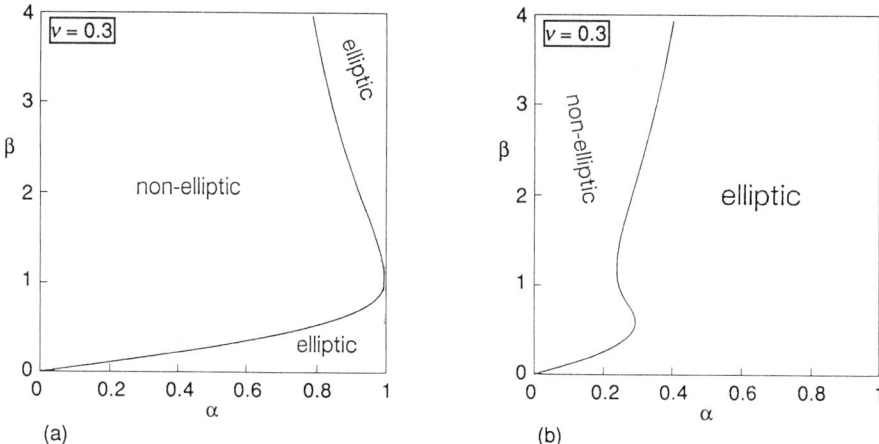

Fig. 10 Critical values of α and β under uniaxial tension (a) and compression (b).

where **n** is the unit vector normal to the plane of discontinuity (band) and **m** is another unit vector. Equation (19) has to be satisfied by **m** and **n** for the localization mode to be possible. The onset of the localization occurs at the first point in the deformation history for which a nontrivial solution of Equation (19) exists.

For the transformation with shear and dilatant effects, such as those in zirconia-containing ceramics (TZP and PSZ), there are two material parameters dominating the constitutive response, i.e., the hardening coefficient α and the factor $A(=3h_0\varepsilon^{pv})$ which governs the influence of shear. The localization analysis of ceramics was first performed by Stam (1994) for plane strain condition. A more general analysis is recently given by Yan et al. (1994b). By using the constitutive model of Sun, Hwang and Yu (1991), Yan et al. (1994b) proved that the critical condition of localization in ceramics for a non-proportional loading history can be expressed as

$$\frac{A^2}{(\sigma_e^M)^2}\left[(n_i n_j s_{ij}^M)^2 - 2(1-v)n_k n_i s_{ij}^M s_{kj}^M\right] - 2(1+v)\frac{A\varepsilon^{pv}}{\sigma_e^M}n_k n_j s_{kj}^M$$
$$+\frac{16-20v}{45}A^2 + \alpha\left[\frac{14-10v}{45}A^2 + 2(1+v)(\varepsilon^{pv})^2\right] = 0 \quad (20)$$

where v is the Poisson's ratio.

For the case of pure dilational transformation ($A = 0$), it can be shown that the localization develops (ellipticity is lost) when $\alpha = 0$, i.e., the slope $B(=\dot{\Sigma}_m/\dot{E}_{ii})$ of the volume stress-volume strain curve ($\Sigma_m \sim E_{ii}$) has a negative value $-4\mu/3$ (μ is the elastic shear modulus). This theoretical prediction is in agreement with the critical transformation defined by Budiansky et al.(1983).

The localization conditions for uniaxial tension and compression are calculated respectively by Yan et al. (1994b, see Figures 10 and 11), and the result indicates that for the Ce-TZP ceramics under uniaxial tension stress state the localized deformation band appears when $\alpha \le 1$. For $A = 3\varepsilon^{pv}$ (or $h_0 = 1$) the plane of the band is normal to the tensile axis (Figure 11), which is in agreement with the available experimental observations (Qing et al., 1993; Sun, Zhao, Chen, Qing and Dai, 1994).

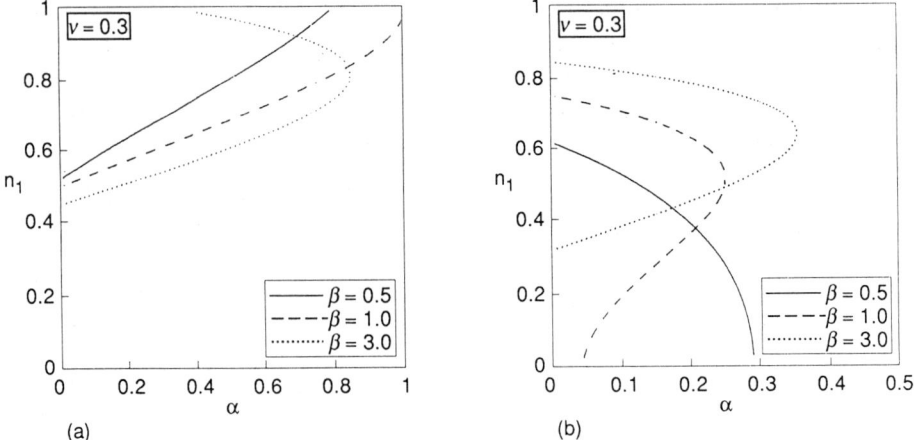

Fig. 11 The variation of the band orientation with α and β, (a) uniaxial tension and (b) uniaxial compression.

For localization under uniaxial compression, acceptable agreement between theoretical prediction and experiments (Reyes-More and Chen, 1988) is obtained. The implications of localization on the material toughening and microstructural design will be illustrated in the following.

5. Transformation Toughening and Microstructure Design - Some new Advances

5.1. Transformation toughening

The discovery of transformation toughening phenomena by Garvie et al. (1975) caused an explosion of research work in the area of zirconia-reinforced ceramic composites. Since the publication of that milestone work, significant progress has been made in improving the mode-I fracture toughness of these materials. Now it has been commonly accepted that stress induced martensitic transformations can substantially enhance the fracture toughness of brittle solids (Evans and Cannon,1986). The effect has been most extensively investigated in ZrO_2 and in various matrices with a ZrO_2 second phase. In this system, the transformation entails a change from the tetragonal (t) to the monoclinic (m) phase and the toughness of certain ceramics can be increased, sometimes by an order of magnitude, by transformation plasticity at the crack tip region (Evans,1990). Such large effects are unprecedented in material systems incapable of crack tip blunting. The scientific basis underlying this phenomenon has progressed at a similarly remarkable pace by contributions from the ceramic science and applied mechanics. The fundamental basis for comprehending the toughening phenomenon resides in the thermodynamics and associated kinetics of the stress induced transformation. The constitutive laws governing stress-induced martensitic transformations lead directly to descriptions of crack tip fields which, in turn, define the levels of transformation toughening. The toughening effect can be modelled in two ways. In the first approach, the stress-shielding effect at the crack tip due to the residual strain fields which develops following transformation, is computed using the constitutive law and fracture mechanics. This method was initially proposed by McMeeking and Evans (1982) and Budiansky et al.(1983) and was later refined by other researchers. The second approach is to directly estimate the energy dissipation due to transformation that occurs as the crack advances. Budiansky et al.(1983) have performed these calculations for a series of dilatant transforming materials with different strain hardening or softening, and this approach can be extended to the transformations with dilatant and shear effects. For a systematic description of the associated scientific principles and their applications, the readers are referred to papers by Evans and Cannon (1986) and Evans (1990).

Recently, anomalous crack tip transformation zone morphology resulting from autocatalysis transformation attracts strong interest of the researchers in this area. Since such anomalous frontal zones will decrease the degree of shielding and thus be detrimental to toughening, the scientific understanding and simulation of this phenomenon, and further control of the spreading of the localized transformation zone by microstructure design represents a significant research challenge in this area. A recent study is concentrated on two aspects: (1) the role of autocatalysis and shear effects on plasticity and toughening, (2) optimum microstructure design.

5.2. Some experimental phenomena

Recent experimental studies on stress-strain behaviours and transformation zones of zirconia ceramics have shown an interesting anomaly (Chen and Reyes-Morel, 1986; Reyes-Morel and Chen, 1988; Rose and Swain, 1988; Yu and Shetty, 1989; Yu *et al.*, 1992; Sun, Zhao, Chen, Qing and Dai, 1994; Tsai *et al.*, 1991; Marshall, 1990; 1992): In magnesia-partially-stabilized zirconia (Mg-PSZ) ceramics the transformation zone fronts ahead of the crack are found to be nearly semicircular, and the corresponding stress-strain curves in both tension and compression show a gradual transition from elastic to plastic behaviour. The ceria-stabilized tetragonal zirconia polycrystals (Ce-TZP) ceramics on the other hand, however, exhibit typically thin elongated crack tip zones with zone lengths approximately 10 times the zone widths (see Figure 12), and the stress-strain curve exhibits burst transformation. A distinct yield point (load drop) and localized transformation bands are observed in tension, compression and bending. Some investigations (Reyes-Morel and Chen,1988; Tsai *et al.*, 1991) have demonstrated that the typical behaviour in Ce-TZP is resulting from the shear effect and autocatalytic transformation. Thus a potential link between shear effect, autocatalytic transformation and the elongated zone shape and localized band in Ce-TZP has been speculated. Recently a lamellar composite containing alternating layers of Ce-TZP and a mixture of Al_2O_3 and $Ce-ZrO_2$ has been fabricated using a colloidal technique and the following encouraging results have been obtained from experiments (Marshall *et al.*, 1991; Marshall, 1992) (Figure 13): (1) The layers interacted strongly with the transformation zones surrounding cracks and indentations, causing the zones to spread along the regions adjacent to the layers and

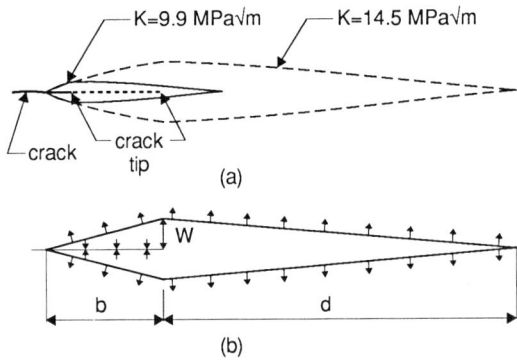

Fig. 12 (A) Transformation zone boundaries in Ce-TZP ceramics during autocatalytic transformation corresponding to two stages of crack growth (Marshall,1990) : solid lines at crack extension of 180 μm, broken lines at crack extension of 1.3 mm. (B) Zone shape and tractions used to calculate crack-tip shielding for the zones of (A).

leading to enhanced fracture toughness. (2) This multilayered microstructure exhibited R-curve behaviour for cracks oriented normal to the layers, with the critical stress intensity factor increasing by a factor of 3.5 from the starting toughness of the Ce-TZP (~5MPa$\sqrt{m}$) to a value of at least 17.5 MPa$\sqrt{m}$ (this value is not saturated to a steady state). (3) Zone spreading and toughening effects were observed for cracks growing parallel to the layers as well as for those oriented normal to the layers. Thus the understanding and simulations of the above mentioned phenomenon by a proper theoretical model and at the same time further employing the phenomenon in microstrucural design of advanced materials represent a very challenging research subject. From the continuum mechanics points of view, we need to know why autocatalytic transformation leads to the above mentioned phenomena, and how to incorporate such underlying mechanism into the constitutive description and how to predict and

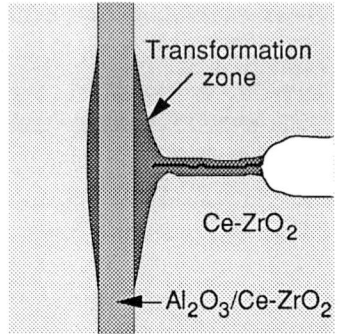

Fig. 13 Transformation localization zone propagating perpendicular to the crack when a mesostructural Al_2O_3/Ce-ZrO_2 plate is laid in front of the crack.

simulate the phenomena by theoretical analysis and numerical computations. Since such experimental phenomena commonly exist in many kinds of phase transitions, the investigation on this subject is undoubtedly of great theoretical and practical importance. The recent advances are reported below.

5.3. The effects of transformation shear and autocatalysis on toughening

Shear effects on transformation toughening have been explored by some researchers from theoretical, numerical and experimental points of view (Lambropoulos, 1986; Chen and Reyes-Morel, 1986; Reyes-Morel and Chen, 1988; Sun, Hwang and Yu, 1991; Stam, 1994; Stam et al., 1994, etc.) and are found to be very substantial. Recent research interest in this aspect is on the effect of shear on toughening and the crack tip transformation zone morphology (see Stam (1994); Stump (1991) and Stump and Budiansky (1990)). The recent numerical study for the transformation zone and toughening under both small scale transformation condition and single edge notched beam specimen presented by Guo et al. (1994) demonstrates: (1) The transformation zone shape changes abruptly when the localization condition of transformation is approached. This is characterized by two important parameters α and h_0 which reflect the contribution of autocatalysis and transformation shear, respectively. The profound effect of autocatalysis and shear is well incorporated by the micromechanics constitutive model of Sun, Hwang and Yu (1991). (2) By properly selecting the parameters α and h_0, the observed narrow-elongated and the branching crack-tip transformation zones of Ce-TZP at low temperatures, where autocalysis prevails, can be successfully simulated in the finite element program by using this model (see Figure 14). Thus the phenomenon of the anomalous crack-tip zone morphology of Ce-TZP observed in experiments (Rose and Swain, 1988; Yu and Shetty, 1989) is explained from the continuum mechanics point of view. (3) The negative shielding effect of transformation toughening for a stationary crack under plane stress condition is discovered first (Figure 15). The toughness enhancement ΔK^{tip} values are calculated for various α and h_0. The results agree with the J-integral calculations which are no longer path independent. Since the "negative shielding effect" generally exists for a stationary crack with all parameters of α and h_0 under plane stress condition, it is not very clear whether such an effect also exists in the case of plane strain. So a further calculation will be performed to check this point. The crack tip frontal zone simulation and toughening calculation for growing crack (the R-curve) are in progress.

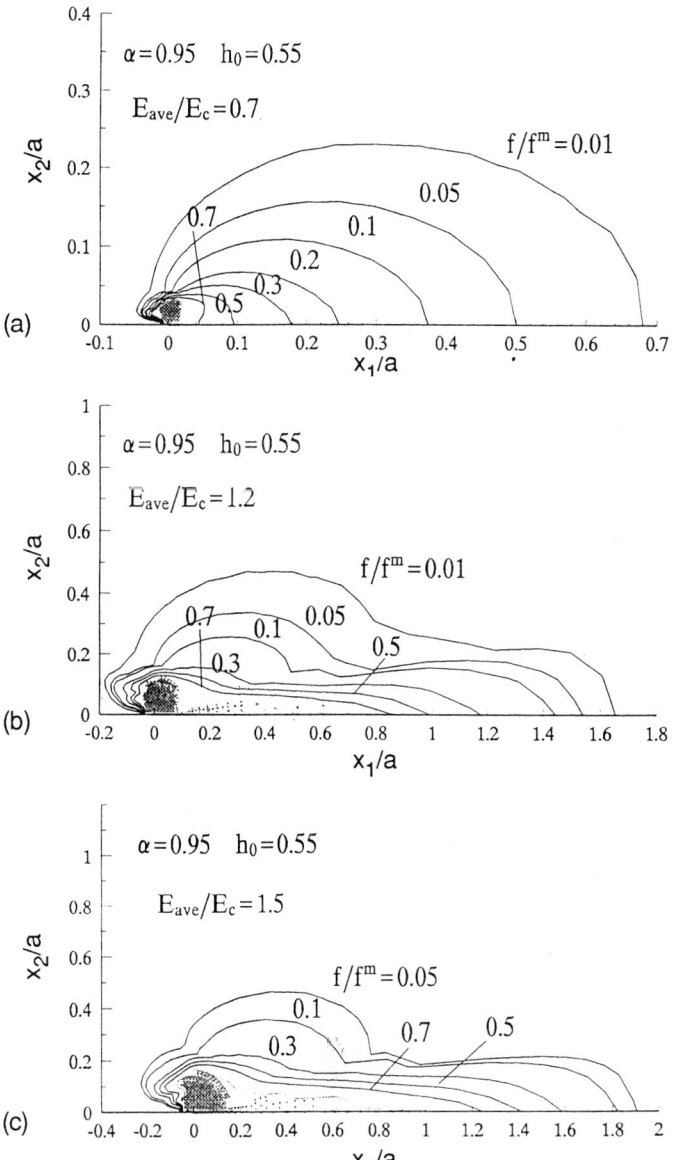

Fig. 14 Numerical results of the typical crack tip transformation zone shape due to autocatalytic transformation by using the model of Sun, Hwang and Yu(1991a): the variation of the transformation zone shape with the applied load (where E_{ave} is the average tensile strain in the specimen calculated by the applied node displacement, E_c is the tensile yielding strain) for a stationary crack in the case of $h_0 = 0.55$, $\alpha = 0.95$ (a,b,c).

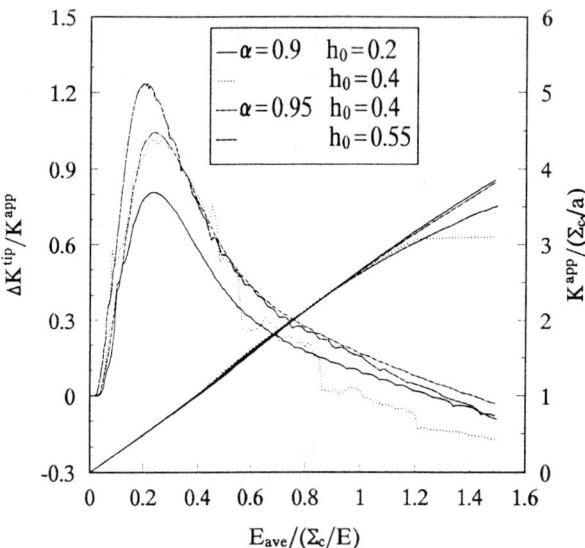

Fig. 15 The variation in the amount of shielding (curves with peaks) and in the calculated applied stress intensity factor (monotonically increasing curves) with the applied displacement loading. E — Young's modulus, a — crack length, E_c, E_{ave} — same as in Figure 14. Stationary crack in the cases of α =0.9, h_0 = 0.2, 0.4, and α =0.95, h_0 =0.4, 0.55 . (Guo et al., 1994).

5.4. Some advances on optimum material microstructure design

Advanced composites, due to their combined advantage in various physical and mechanical properties, are finding ever increasing applications in areas of modern high technology. The development of such material is an essential concern in the building of space platforms, hypersonic vehicles and other military systems. Many of these applications require materials which will survive against the impact loading and have the high capacity of energy absorption, high strength and toughness and resilience after inelastic deformation. In developing such materials, modelling their properties and predicting their behaviour poses challenging theoretical problems to the scientists of both materials and mechanics by the fact that these composites are often manufactured and synthesized under the guidance of material design. They are generally constituted by two or more materials with unequal elastic constants and stress-strain responses, *etc.*. They develop highly inhomogeneous stress and strain fields when loaded either thermally or mechanically. The characteristic length scale of the inhomogeneity of the stresses and strains is determined by the geometry of the microstructure. Successful fundamental research into the properties and reliability of the composite therefore requires theoretical models that treat the materials as discontinuous on the same microscopic scale and predict the macroscopic behaviour correctly. In doing this, micromechanics approach will play a dominant role.

(1) Effect of dual-scale microstructure on toughness of laminar zirconia composite. The numerical simulation by Sun, Guo and Li (1994) successfully reproduced the experimentally observed two effects of the dual-scale microstructure (see Figure 13) on the toughness of a

lamellar composite, *i.e.*, the truncation of the elongated frontal zone that forms in single phase Ce-ZrO_2 and the spreading of the transformation zone along the layers normal to the crack plane. Quantitative analysis on the role of microstructure in transformation toughening is first carried out in their work which will provide a starting point for the microstructural design in the future.

(2) Micromechanics analysis on the microstructure design of two-phase pseudoelastic composite. Recently, a micromechanics analysis and scientific foundation for the possibility of designing a two-phase pseudoelastic composite with pseudoelastic toughening have been formulated for the case where ductile transformable shape memory alloy particles are embedded coherently in an elastic matrix. It is demonstrated that a pseudoelastic stress-strain loop in a macroscopic loading-unloading cycle can be obtained by microscopically stress induced forward and reverse martensitic transformations in the shape memory alloy particles. The toughening analysis of this innovative microstructure is given and a theoretical formula for the evaluation of toughness enhancement due to the pseudoelasticity is derived (Sun, 1994). A numerical study of this pseudoelastic toughening was recently performed by Stam (1994).

Acknowledgements

The authors are grateful to the National Natural Science Foundation and the State Commission of Education of China for financial support.

References

Abeyaratne, R. (1993), Material instability and phase transitions in thermoelasticity. In: *Theoretical and Applied Mechanics 1992* (eds. S.R. Bodner *et al.*), Elsevier, Amsterdam, 53.

Abeyaratne, R., S.-J. Kim and J.K. Knowles (1993), A continuum model of a thermoelastic solid capable of undergoing phase transitions, *J. Mech. Phys. Solids* **41**, 541.

Budiansky, B., J.W. Hutchinson and J.C. Lambropoulos (1983), Continuum theory of dilatant transformation toughening in ceramics, *Int. J. Solids Struct.* **19**(3), 337.

Chen, I.W. and P.E. Reyes-Morel (1986), Implications of transformation plasticity in zirconia-containing ceramics: I, shear and dilation effects, *J. Am. Ceram. Soc.* **69**(3), 181.

Christian, J.W. (1975), *The Theory of Transformations in Metals and Alloys*, Pergamon, Oxford, 2nd ed., Vol.1.

Delaey, L., R.V. Krishnan, H. Tas and H. Walimont (1974), Review thermoelasticity, pseudoelasticity and the memory effects associated with martensitic transformations, *J. Mater. Sci.* **9**(9),1521.

Ericksen, J.L. (1975), Equilibrium of bars, *J. Elasticity* **5**, 191.

Eshelby, J.D. (1956), The continuum theory of lattice defects. In: *Solid State Physics* (eds. F.Seitz and D.Turnbull) ,Academic Press, New York, Vol.3, p. 79.

Eshelby, J.D. (1970), Energy relations and the energy momentum tensor in continuum mechanics. In: *Inelastic Behaviour of Solids* (eds. M.F.Kanninen *et al.*), McGraw-Hill, New York, p. 77.

Eshelby, J.D. (1957), The determination of the elastic field of an ellipsoidal inclusion and related problems, *Proc. R. Soc. London A* **241**, 376.

Evans, A.G. and R.M. Cannon (1986), Toughening of brittle solids by martensitic transformations, *Acta Metall.* **34**(5), 651.

Evans, A.G. (1990), Perspective on the development of high-toughness ceramics, *J.Am. Ceram. Soc.* **73**(2), 187.

Falk, F. (1980), Model free energy, mechanics and thermodynamics of shape memory alloys , *Acta Metall.* **28**, 1773.

Falk, F. (1982), Landau theory and martensitic phase transitions, *J. de Phys.*, Colloque C4, supplement to 43(12), C4-3.

Falk, F. and P. Konopka (1990), Three-dimensional Landau theory describing the martensitic phase transformation of shape memory alloys, *J. Phys.: Condens. Matter* **2**, 61.

Garvie, R.C., R.H.J. Hannink and R.T. Pascoe (1975), Ceramic Steel? *Nature* **258**, 703.

Guo, T.F., Q.P. Sun and X. Zhang (1994), The role of autocatalysis and transformation shear in crack tip zone shape and toughening of zirconia ceramics, submitted to *Int. J. Solids Struct.*

Hwang, K.C. (1989), *Nonlinear Continuum Mechanics* (in Chinese), Tsinghua University Press, Beijing.

James, R.D. (1986), Displacive phase transformations in solids, *J. Mech. Phys. Solids* **34**, 359.

Lambropoulos, J.C. (1986), Shear, shape and orientation effects in transformation toughening, *Int. J. Solids Struct.* **22**(10), 1083.

Marshall, D.B. (1992), Design of high-toughness laminar zirconia composite, *Ceram. Bull.* **71**(6), 969.

Marshall, D.B. (1990), Crack shielding in ceria-partially-stabilised zirconia, *J. Am.Ceram.Soc.* **73**(10), 319.

Marshall, D.B., J.T. Ratto and F.F. Lange (1991), Enhanced fracture toughness in layered microcomposites of Ce-ZrO_2 and Al_2O_3, *J. Am. Ceram. Soc.* **74**(12), 2979.

McMeeking, R.M. and A.G. Evans (1982), Mechanisms of transformation toughening in brittle materials, *J. Am. Ceram. Soc.* **65**(5), 242.

Mori, T. and K. Tanaka (1973), Average stress in matrix and average elastic energy of materials with misfitting inclusions, *Acta Metall.* **21**(5), 571.

Mura, T. (1987), *Micromechanics of Defects in Solids*, Martinus Nijhoff, The Hague, The Netherlands.

Mura, T. (1988), Inclusion problems, *Appl. Mech. Rev.* **41**(1), 15-20.

Müller, I. (1989), On the size of the hysteresis in pseudoelasticity, *Continuum Mechanics and Thermodynamics* **1**, 125-142.

Müller, I., and H. Xu (1991), On the pseudoelastic hysteresis, *Acta Metall.* **39**, 263.

Patoor, E., A. Eberhardt and M. Berveiller (1987), Potentiel pseudoelastique et plasticite de transformation martensitique dans les monoet polycristaux metalliques. *Acta Metall.* **35**, 2779.

Patoor, E., A. Eberhardt and M. Berveiller (1988), Thermomechanical behaviour of shape memory alloys, *Arch. Mech.* **40**, 775.

Qing, X.L., Q.P. Sun, and F.L. Dai (1993), Study of transformation plasticity in tetragonal zirconia polycrystals by moiré interferometry, *Acta Mechanica Sinica* **9**(4), 330.

Rao, C.N.R. and K.J. Rao (1978), *Phase Transitions in Solids*, McGraw-Hill, New York.

Reyes-Morel, P.E. and I.W. Chen (1988), Transformation plasticity of ceria-stabilized tetragonal zirconia polycrystals: I, stress assistance and autocatalysis, *J. Am. Ceram. Soc.* **72**(5), 343.

Reyes-Morel, P.E., J.-S. Cherng and I.W. Chen (1988), Transformation plasticity of ceria-stabilized tetragonal zirconia polycrystals: II, pseudoelasticity and shape memory effect, *J. Am. Ceram. Soc.* **72**(8), 648.

Rice, J.R. (1971), Inelastic constitutive relations for solids: an internal-variable theory and its application to metal plasticity, *J.Mech. Phys. Solids* **19**, 433.

Rice, J.R. (1975), Continuum mechanics and thermodynamics of plasticity in relation to microscale deformation mechanisms. In: *Constitutive Equations in Plasticity* (ed. A.S. Argon), MIT Press, Cambridge, MA, p. 23.

Rice, J.R. (1976), The Localization of plastic deformation. In: *Theoretical and Applied Mechanics* (ed. W.T. Koiter), North Holland, Amsterdam, p. 207.

Rose, L.R.F. and M.V. Swain (1988), Transformation zone shape in ceria-partially-stabilized zirconia, *Acta Metall.* **36**(4), 955.

Song, G.Q. (1994), *A constitutive model of martensitic transformation by combined Landau-Devonshire theory and micromechanics approach* (in Chinese), Ph.D. Thesis, Dept. of Eng. Mech., Tsinghua University, Beijing, China.

Stam, G.Th.M. (1994), *A micromechanical approach to transformation toughening in ceramics*, Ph. D. Thesis, Delft University of Technology, Delft, The Netherlands.

Stam, G.Th.M., E.van der Giessen and P. Meijers (1994), Effect of transformation-induced shear strains on crack growth in zirconia-containing ceramics, *Int. J. Solids Struct.* **31**(14), 1923.

Stump, D.M. (1991), The role of shear stresses and shear strains in transformation Toughening, *Phil. Mag. A* **64**(4), 879.

Stump, D.M. and B. Budiansky (1990), Crack growth resistance in transformation toughened ceramics, *Int. J. Solids Structure* **25**, 635.

Sun, Q.P.(1989), *Micromechanics Constitutive Theory of Transformation Plasticity and Toughening Analysis*, Ph.D. Thesis, Tsinghua University, Beijing, China.

Sun, Q.P. (1994), A micromechanics analysis for the microstructure design of two-phase pseudoelastic composite, *Acta Mech. Sinica* **10**(2),162.

Sun, Q.P., T.F. Guo and X.J. Li (1994), Effect of dual-scale microstructure on the toughness of laminar zirconia composite, submitted to *Int. J. of Fracture*.

Sun, Q.P., Y. Huang, S.W. Yu and K.C. Hwang (1991), Toughening analysis of mode III crack in transformation toughened ceramics, *Acta Mechanica Solida Sinica* **4**(3), 251.

Sun, Q.P., Y. Huang, S.W. Yu and K.C. Hwang (1990), Experimental and numerical research on transformation plasticity and toughening of Ce-TZP ceramics. In *Proceedings of C-MRS International Symposia* (eds. D.S. Yan et al.), Beijing, China, Elsevier Science Publishers, North-Holland, Vol.5, p. 195.

Sun, Q.P. and K.C. Hwang (1991), Micromechanics modeling for the constitutive behaviour of polycrystalline shape memory alloys. In: *Proceedings of the IUTAM Symposium on Constitutive Relations of Finite Deformaiton of Metals* (eds. R.Wang and D.C.Drucker), Springer-Verlag, Peking University Press, Beijing, China, p. 188.

Sun, Q.P. and K.C. Hwang (1992), The crack tip zone shielding effect for the ductile particle reinforced brittle materials, *Acta Mechanica Sinica* **8**(2), 136.

Sun, Q.P. and K.C. Hwang (1993a), Micromechanics modelling for the constitutive behaviour of polycrystalline shape memory alloys, I: derivation of general relations, *J. Mech. Phys. Solids* **41**(1), 1.

Sun, Q.P. and K.C. Hwang (1993b), Micromechanics modelling for the constitutive behaviour of polycrystalline shape memory alloys, II: study of the individual phenomena, *J. Mech. Phys. Solids* **41**(1), 19.

Sun, Q.P. and K.C. Hwang (1994), Micromechanics constitutive description of thermoelastic martensitic transformation. In: *Advances in Applied Mechanics* (eds. J.W.Hutchinson and T.Y.Wu), Academic Press, New York, Vol.31, p. 249.

Sun, Q.P., K.C. Hwang, and S.W. Yu (1991), A micromechanics constitutive model of transformation plasticity with shear and dilatation effects, *J. Mech. Phys. Solids* **39**(4), 507.

Sun, Q.P., Y.W. Yan and K.C. Hwang (1994), A generalized micromechanics constitutive theory of single crystal with thermoelastic martensitic transformation, submitted to *J. Mech. Phys. Solids*.

Sun, Q.P., S.W.Yu and K.C.Hwang (1990), A micromechanics constitutive model of pure dilatational martensitic transformation of ZrO_2-containing ceramics, *Acta Mechanica Sinica* **6**(2),141.

Sun, Q.P., Z.J. Zhao, W.Z. Chen, X.L. Qing, X.J. Xu and F.L. Dai (1994), Experimental study of stress-induced localized transformation plastic zones in tetragonal zirconia polycrystalline ceramics, *J. Am. Ceram. Soc.* **74**(5), 1352.

Tsai, J.F., C-S. Yu and D.K. Shetty (1991), Role of autocatalytic transformation in zone shape and toughening of ceria-tetragonal-zirconia-alumina(Ce-TZP/ Al_2O_3) composites, *J. Am. Ceram. Soc.* **74**(3), 678.

Wayman, C.M. (1964), *Introduction to the Crystallography of Martensitic Transformation*, Macmillan, New York.

Wechsler, M.S., D.S. Lieberman, and T.A. Read (1953), On the theory of formation of martensite, *Trans., AIME* **197**, 1503.

Yan, W.Y. (1994), *Micromechanics Constitutive Study of Transforming Single Crystals* (in Chinese), Ph.D. Thesis, Dept. of Eng. Mechanics, Tsinghua University, Beijing, China.

Yan, W.Y., Q.P. Sun and K.C. Hwang (1994a), A micromechanics constitutive model of single crystal with thermoelastic martensitic transformation, submitted to *Acta Metall. Mater.*.

Yan, W.Y., Q.P. Sun and K.C. Hwang (1994b), Analyses of transformation plastic localization in ceramics, submitted to *Int. J. Solids Struct.*

Yu, C-S. and D.K. Shetty (1989), Transformation zone shape, size and crack-growth-resistance (R-curve) behaviour of ceria-partially-stabilized zirconia polycrystals, *J.Am.Ceram.Soc.* **72**(6), 921.

Yu, C-S., D.K. Shetty, M.C. Shaw and D.B. Marshall (1992), Transformation zone shape effects on crack shielding in ceria-partially-stabilized zirconia (Ce-TZP)-alumina composites, *J. Am. Ceram. Soc.* **75**(11), 2991.

J.G.M. Van Mier, E. Schlangen*, A. Vervuurt* and
M.R.A. Van Vliet**

Damage Analysis of Brittle Disordered Materials: Concrete and Rock

Reference: Van Mier, J.G.M, E. Schlangen, A. Vervuurt and M.R.A. van Vliet (1995), Damage Analysis of Brittle Disordered Materials: Concrete and Rock. In: *Mechanical Behaviour of Materials* (ed. A. Bakker), Delft University Press, Delft, The Netherlands, pp. 101-126.

Abstract: Results of a combined experimental and numerical study on crack processes in concrete and rock are presented. The fracture process in these brittle disordered materials is a complicated non-uniform process ranging from distributed microcracking to crack face bridging. Crack-face bridging has been identified as the main toughening mechanism leading to the long tail in the softening diagram. The crack-face bridges are overlapping crack tips near stiff inclusions in the material. Because of the non-uniformity of the fracture process, the rotational stiffness of the loading platens has a significant effect on the fracture energy and stress-crack opening behaviour in tension. The boundary effects were analysed using the numerical latttice model and compared with experimental observations.

In the lattice model the material is discretized in a network of beam elements. After the adjustment of the elastic stiffness of the lattice, fracture can be simulated quite realistically. In the paper different approaches are compared. The results suggest that realistic fracture patterns can be simulated when the material structure is carefully mapped on a regular or random triangular lattice. The fracture law at the beam level in the lattice model is still open to debate, as is the method needed for the determination of the model parameters. These issues are also addressed in this paper.

1. Introduction

Fracture mechanics of concrete is a rapidly expanding field of research. The effort is concentrated on deriving numerical models to analyse the behaviour of concrete structures. It is essential in such approaches that the parameters needed in the numerical models can be derived in a simple and straightforward manner. It would be most convenient if the material properties could be measured directly from fundamental experiments in tension, (multiaxial) compression or shear. Complex interactions exist however, between the fracture zone, the specimen geometry (size and shape) and boundary conditions. This implies that it is not possible to retrieve material properties directly from experiments. Instead a (semi-) inverse modelling technique should be used to derive the input parameters for any given model. Obviously a relationship exists between the model representing the phenomena that are simulated, and the required input parameters. The physical basis for such a model is important, as it is hoped this will eventually lead to well-defined material constants. At present the most simple model containing the least elaborate parameters seems the best choice for concrete and rock, provided of course that the model is capable of capturing the most essential fracture phenomena in these materials.

Concrete, rock and (non-transformable) ceramics are complicated brittle matrix composites. The macroscopic or global fracture behaviour of such materials is highly non-linear. In fact the global response represents structural changes occuring at the local level, i.e. the microscopic or mesoscopic material structure. For concrete the typical size range of the material structure responsible for the macroscopic mechanical behaviour is the meso-level, where aggregate particles are embedded in a

---

* *Delft University of Technology, Department of Civil Engineering,Stevin Laboratory, P.O. Box 5048, 2600GA Delft, The Netherlands.*

brittle cement matrix. The typical size range of the aggregates is between several hundreds of a micrometre up to 20 mm. For ceramics the size range is much smaller, i.e. it is confined to the micrometre range, whereas for rocks microstructures are found at a variety of size scales. For example, sandstones are found with a rather uniform particle structure in the 1-2 mm range, but there are also conglomerates that resemble concrete with particles sizes up to 50 mm.

An essential feature of fracture of the aforementioned brittle disordered materials is the fact that microcracking occurs at the level of the material structure and preceeds the development of global cracks. Basically, it seems that mode I fracturing occurs before frictional slip takes place. Even under global compressive stress, tensile microcracking is observed; it precedes the growth of global shear fractures. Recently developed lattice-type fracture models are an interesting technique to capture these basic microcrack phenomena. In lattice models the continuum is discretized in a network of bar or beam elements. The properties of the lattice elements have to be chosen such that the overall elastic properties of a complete lattice resemble the global elastic properties of the material that is modelled. Fracture can be simulated in a lattice by removing a lattice element after an effective stress in the element exceeds the strength that was assigned to that particular element. The assignment of strength and stiffness to the lattice elements can be done in a variety of manners. Perhaps the simplest way is the selection of an appropriate stochastic distribution of beam strengths and the assignment of different values to each of the lattice elements. Another approach would be to map the material structure, either a computer-generated structure or a digital image taken from a section through the real material, directly on top of a lattice.

In this paper different types of lattice models for simulating fracture in brittle disordered materials will be outlined. The emphasis will be on the discretization method, the procedure to introduce heterogeneity, the definition of a fracture law at the level of the lattice elements, and the parameter identification procedure. At several stages examples of simulations of laboratory scale specimens will be presented and, where possible, be compared with experimental results. Many experiments were performed parallel to the simulations. The main goal of these exercises was to come to an improved fundamental insight into the fracture processes in brittle disordered materials. The simulation technique gives promising results so far, in particular for global tension and combined tension/shear. Efforts are being undertaken to modify the models so that compression can be simulated as well. The technique may possibly be suitable for use in the near future for engineering materials with carefully designed properties for specific applications. Moreover, as will be demonstrated, the technique is a useful tool in the search for reliable laboratory procedures.

2. Mechanisms of Fracture in Concrete and Rock

Before addressing the basics of lattice-type models the existing ideas on fracture mechanisms in concrete and sandstone will be briefly described. The discussion will be limited to these materials as they have some elements of behaviour in common. Let us consider the particle model of Figure 1. The load transfer and fracture mechanisms under tension and compression are drawn schematically. As mentioned, these models are thought to be representative of cohesive particle models such as concrete and sandstone, where the particles are bonded together through a brittle matrix. The difference between concrete and sandstone seems to be the size of the aggregates: they are quite uniformly distributed in, for example, Colton or Felser sandstone, whereas a complete hierarchy of particle sizes appears in concrete. In the model described here it is assumed that the aggregates have a higher Young's modulus than the surrounding matrix and that the interface between the matrix and the aggregates is the weakest link in the system.

In tension (Figure 1a) fracture of the composite is initiated by microcracking in the transition zone between matrix and aggregate, which is the weakest link in the material; see, for example, Mindess (1989). The interaction between the various aggregate particles is not very important in uniaxial tension, since the entire matrix is subjected to tensile stress. In concrete the interface is rather porous, as hydration of the cement is retarded; see, for example, Bentz et al. (1992). The situation is more complicated, however, as premature debonding may occur. Shrinkage cracks may develop around the aggregates due to differential temperature and humidity in the hardening concrete or due to bleeding. The shrinkage cracks develop in most cases in preferential directions, causing initial anisotropy of the material. As a result different properties can be found when the tensile loading direction is changed; see, for example, Hughes & Ash (1970).

In compression, on the other hand, the particle interaction is essential in order to generate tensile stresses in the microstructure of the material. In Figure 1b this is shown schematically for a configuration of four circular inclusions. Splitting tensile forces develop, causing vertical tensile cracks, again preferentially in the interfacial zone. The vertical cracks are, however, rather stable in a compressive stress field, and propagation will occur only under further increasing external load. Wing cracks growing from inclined slits in a homogeneous material also stabilize in a compressive stress field; see, for example, Horii & Nemat-Nasser (1985). The particle interactions are essential, not only for concrete; the same mechanism seems to occur in sandstones (Nihei et al. (1994) and Kemeny & Cook (1991)). However, sandstones are generally more porous and there is some debate as to whether the particle interaction is the governing mechanism or if perhaps the porosity plays a more important role. However, both mechanisms lead to axial cleavage cracking at moderate levels of global stress. A secondary mechanism seems necessary to cause global failure. In many compressive tests on concrete, cone-like mortar elements have been found on top of stiff aggregate particles; see, for example, Vile (1968), Stroeven (1973), Van Mier (1984) and Vonk (1992). These cones suggest that some form of shear fracture must occur in later stages of loading. In Figure 1b some of these failure cones have been drawn, and a shear crack has been schematized as an array of parallel splitting cracks. This "shear crack mechanism" was speculated upon by various researchers, for example Vile (1968). The cones can develop since frictional restraint

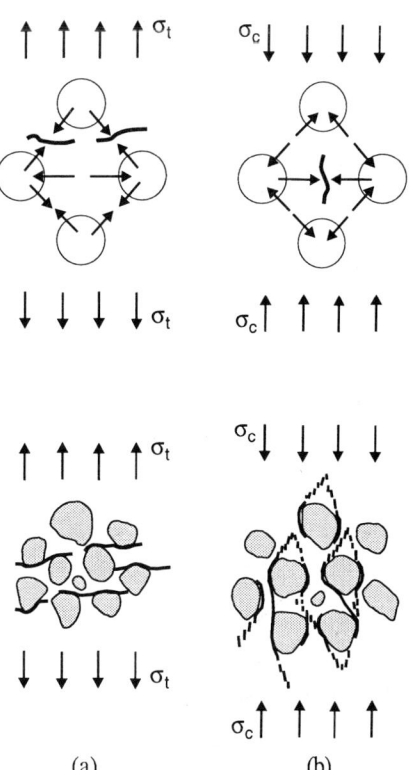

Fig. 1. Micromechanisms of failure in particle composites subjected to tension (a) and compression (b); from Van Vliet & Van Mier (1994).

prevents the 'soft' matrix from 'flow' along the top and bottom of the aggregate particles. The shear stresses that can develop at the interface are shown schematically in Figure 2a; they cause a triaxial compressive state of stress to develop in the cones. In fact this situation is quite similar to the development of triaxially stressed areas below rigid steel platens that are in direct contact with the specimen in an uniaxial compression experiment, as shown in Figure 2b. The main difference is of course that the scale is different. It will be obvious that the triaxial state of stress in the cones on top of and below

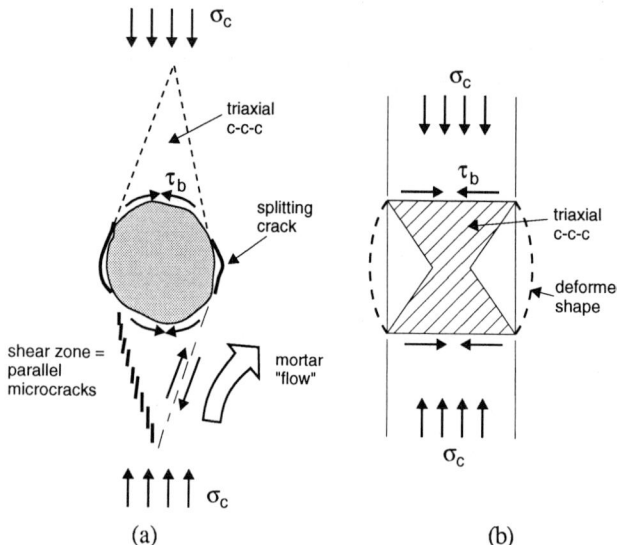

Fig. 2 Triaxially stressed areas (triaxial compression) on top and below the aggregates in concrete (a), and in a concrete specimen tested between rigid loading platens (b); from Van Vliet & Van Mier (1994).

the aggregate particles depends on the ratio of aggregate stiffness to matrix stiffness. When, for example, lightweight (porous) aggregates are used, the state of stress around the particles changes dramatically. In lightweight concrete (containing porous (low strength) aggregates) fracturing seems to initiate in the porous grains and not in the interface as discussed before.

Let us now consider the global experiments. The strength and fracture response of the global specimen in uniaxial compression depends very much on the type of loading platen used in the experiment, as shown by Kotsovos (1983) and confirmed by Vonk et al. (1989), and also on the size (or rather slenderness) of the specimen, Van Mier (1986a). These effects are caused by the aforementioned triaxially stressed zones in the top and bottom parts of a specimen. The amount of boundary restraint (τ_b in Figure 2b) will determine directly the size of the triaxially stressed zone and therefore also the failure mode of the specimen. Both effects, boundary restraint and slenderness, are shown schematically in Figure 3. Of course there must be an interaction of both effects, and these

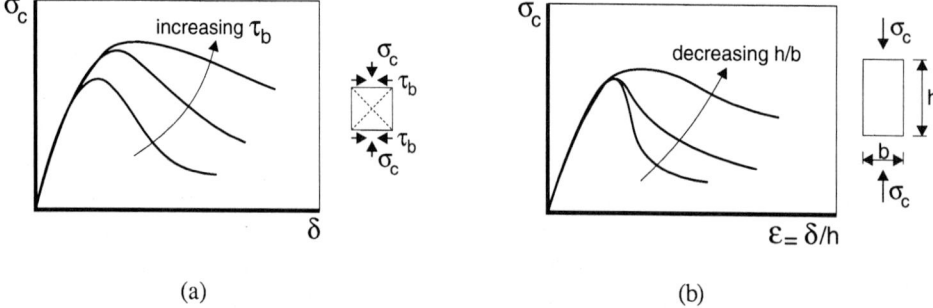

Fig. 3 Effect of boundary restraint (a) and specimen slenderness (b) on the compressive stress-deformation behaviour of concrete; after Van Vliet & Van Mier (1994).

questions are currently under investigation in a round robin test by RILEM technical committee 148SSC (Strain Softening of Concrete). The conclusion must be that a structural property is measured in a compressive experiment (Van Mier, 1984), rather than a material property.

For tensile fracture the same problem is encountered. The crack growth process, and thus the strength and post-peak behaviour are strongly affected by size and boundary rotations in a tensile experiment; see, for example, Van Mier et al. (1994b). Consequently,

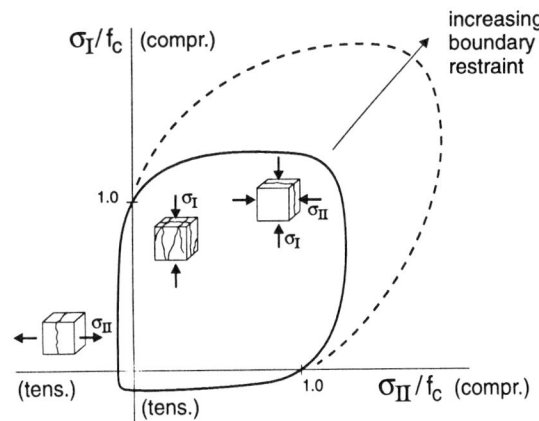

Fig. 4 Biaxial strength envelope (solid line after Kupfer (1973)), and fracture modes. The envelope expands under increasing boundary restraint; after van Vliet & van Mier (1994).

testing alone is not sufficient if we want to determine size- and boundary-independent fracture properties of brittle disordered materials such as concrete and sandstone. The derivation of global fracture properties such as stress-deformation curves, including the post-peak response and the definition of a global fracture energy, may simply be impossible. This still has to be investigated, however.

Again, for multiaxial states of stress the same problems are found as in uniaxial compression. In order not to overestimate the multiaxial compressive strength, complicated test facilities are inevitable; see, for example, Gerstle et al. (1978). When sufficient precautions are taken, biaxial and triaxial failure contours can be determined, indicating the strength of the material under generalized stress. In Figure 4 the often quoted biaxial strength envelope of Kupfer (1973) is shown. The experiments by Kupfer were done using brushes as a load-transmitting medium. Brushes have been shown to be effective in reducing the boundary restraint, especially in the small deformation regime up to peak stress. Figure 4 shows the increase in strength in the biaxial compression regime, and the decrease in strength in the tensile/compressive regimes. Basically the mode of failure in biaxial compression is similar to the failure mode in uniaxial compression, except that the lateral expansion caused by the microcrack growth in the material can occur in the unloaded direction only. When different loading systems are used, for example with higher boundary restraint, the biaxial failure contour expands rapidly, as shown, for example, by Gerstle et al. (1978); see also Figure 4.

The strong boundary- and size-dependent strength and fracture properties of concrete and rocks has led us to develop a simple fracture model at the particle level of the material. *It is essential is that the exact boundary conditions of the experiment are included in the simulations.* Using this model we hope to come to a better understanding of the behaviour of the material under mechanical loading. These efforts may lead to the definition of a laboratory-scale experiment from which macroscopic strength and fracture parameters can be estimated. The success of the lower-scale fracture model depends to a high degree on the parameters used in the model: can they be measured and is a physical interpretation possible? In the next section a number of possible lattice-type fracture models are outlined. Their potential for describing the fracture phenomena in tension and compression is discussed.

3. Lattice-type Fracture Models

3.1. Two Different Approaches

In 1941 Hrennikoff proposed solving problems of elasticity by using discrete truss works. Bar elements were connected by hinges, and the elastic properties of the material were imitated by selecting the appropriate cross-section and elastic constants for the lattice elements. At that time the disadvantage of the truss model was that not sufficient computational power was available to solve large systems. Moreover, since in triangular truss models the Poisson's ratio is always equal to 1/3, only a limited range of materials such as steel and ice can be studied. In the late 1980s lattice models were introduced for simulating fracture; see, for example, Hansen et al. (1989), Herrmann (1991) and Zubelewicz & Bažant (1987). Instead of using trusses, large frameworks were analysed. These have potential for a wider class of materials, as a larger range of Poisson's ratios can be obtained. We were inspired by the development in theoretical physics after the first author attended a summer school in Cargèse (Corsica), where the physicists tried to convince engineers that lattice models were the most appropriate road to solving fracture problems in brittle disordered materials, Charmet et al. (1990). We adopted the principle, but changed the geometry of the initially square lattices proposed by Herrmann (1991) to triangular lattices, Schlangen & van Mier (1992a,b). The advantage of the triangular lattice is the Poisson's ratio, which is obviously zero for a square lattice, when similar elastic constants are specified for all lattice beams. In the model by Herrmann (1991), however, different elements with a Poisson's ratio for each beam were adopted.

A lattice is a discretization of a continuum, and for modelling heterogeneous materials such as concrete, rock and ceramics, disorder has to be introduced in the model. This can be done in various ways. The material structure can be directly projected on top of a lattice; alternatively randomness can be introduced through a distribution of one of the properties of the lattice elements such as length, Young's modulus or fracture strength. Two examples of introducing disorder are described below, namely the lattice with particle overlay and the centre-particle lattice. In addition to these two models a third variant can be found in literature, namely a lattice where each node is not only connected to the neighbouring nodes but where contacts may also exist with nodes that are located at larger distances; see, for example, Burt & Dougill (1977) and Berg & Svensson (1991). Here we will limit ourselves to models where connections exist only between neighbouring nodes. As will be seen, this can be achieved using either regular or random triangular lattices. In a sense the models described here resemble the mechanical models that were presented in Figure 1. Following a description of the models a number of examples will be shown of tensile fracture simulations with different models.

3.1.1. Lattice with particle overlay

The most straightforward approach to introducing disorder is to overlay a lattice on top of a generated particle structure for the material under consideration. In Figure 5 a generated particle structure for concrete is shown with a projected regular triangular lattice. Generating the particle structure in this specific example starts from a three-dimensional Fuller distribution of aggregates. The particles are assumed to be spheres. The number of circular sections in a plane is calculated using

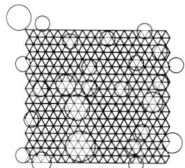 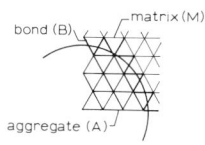 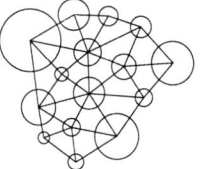

Fig. 5   Lattice projected on top of generated particle structure of concrete; after Schlangen & Van Mier (1992a).

Fig. 6   Centre particle lattice; after Van Vliet & Van Mier (1994)

a formula derived by Walraven (1980). Next, starting with the largest particles, the circular inclusions are placed randomly in a plane. One should of course be careful not to locate aggregates with overlaps. If this occurs, a new trial is made. In most of the analyses that have been done to date, a lower threshold for the aggregate size is specified in order to limit the number of lattice elements. After the particle structure is generated, either a regular triangular or a random triangular lattice (i.e. a lattice with variable beam lengths which is generated from a regular square grid of size s; see Vervuurt et al., 1994) is projected on top of the material structure. Different properties are then assigned to lattice beams falling inside the aggregates, in the matrix material or those beams intersecting with the boundary of an inclusion. In the latter case bond properties are assigned to the beam. This procedure has been used for the last few years and is described in detail in Schlangen & Van Mier (1992a).

The particle structure can be generated following different procedures. In addition to the above method, it is also possible to drop particles from randomly chosen horizontal positions. The particles are randomly selected from a given distribution. Thus again, first a particle distribution is prescribed that resembles a real particle structure of concrete. The examples shown in section 3.3 were generated in this manner. The advantage is that the computational effort is considerably less than with the random placement method. The random 'drop' method will be described in detail in Van Vliet & Van Mier (1995). The placement obtained using this method, however, is not completely random, since holes appear below large aggregates. Horizontal shaking could probably improve the result. Finally, it is possible to use a digital image of a real section through concrete or sandstone as a basis for the lattice model, see Schlangen (1995). Assigning different properties to different beams is then similar to what is described in Figure 5.

It has been argued that the same micromechanical approach can be followed using continuum elements (in 3D analyses) or plane stress elements (in 2D). This has been done by several authors, for example Vonk (1992) and Roelfstra (1989). In many such models interfaces have been made at various locations. Crack growth is then of course limited to these locations. To obtain correct fracture behaviour, however, still complex constitutive laws are used. In the lattice model the fracture law remains very simple indeed, as will be shown in section 3.2.

3.1.2. Centre-particle lattice

An alternative manner of modelling a particle composite with a lattice is the centre-particle lattice. The lattice beams directly represent the interactions between the aggregates, as shown schematically in Figure 1. Again, one can start with generating a material structure as described before, but the lattice is now constructed by connecting the centres of neighbouring particles, as shown in Figure 6. This approach has been followed by several researchers, for example Zubelewicz & Bažant (1987) and Beranek & Hobbelman (1994). The identification of the lattice beam properties now becomes more complicated, as each beam comprises part of an aggregate, the bond zone and part of the matrix. Moreover, since this model is linked to a presumed stress transfer mechanism in the particle composite, it can only be used for such a specific composite. The model described in the previous section seems to have a wider application. The same fracture laws as used for the overlay model can be used, but again in this case the selected fracture law should correspond with the presumed stress transfer between the particles. Note that a lattice constructed in this way resembles a random lattice, i.e. a lattice with different beam lengths.

3.2. Fracture Laws

In all the numerical work carried out to date the aim has been to model the fracture process using the simplest possible fracture law. At present we are in a transitory stage, as the first proposal for the fracture law seems to break down under global compressive load; see also section 6. In the past we have done extensive simulations using the particle overlay method of section 3.1.1, in combination both with a regular triangular lattice and with a lattice with random beam lengths. Since the beginning of our investigations we have used a fracture law based on normal force and bending moment in the lattice beams. The law is based on the maximum tensile stress that occurs in the outermost fibres of a lattice beam due to the normal force F and the bending moments M_i and M_j in the nodes i and j respectively,

$$\sigma_t = \frac{F}{A} + \alpha \frac{(|M_i|,|M_j|)_{max}}{W} \tag{1}$$

where $A = bh$ is the cross-sectional area of the beam and $W = bh^2/6$ the section modulus. When the stress σ_t exceeds the strength assigned to a particular beam, the beam is removed from the mesh and a new computation is made, followed by the next removal. The failure strength of the beams depends on the particular method used for mapping the material structure on the lattice. In section 4 more attention will be given to the parameter identification procedure. Previous research has shown that a value $\alpha = 0.005$ gave realistic results in tensile and shear simulations. This parameter remains rather debatable, however, as a physical basis is lacking. The effect of α will be considered again in the next section.

Recently we have started to use another fracture law proposed earlier by Beranek & Hobbelman (1994), which seems to have a better physical basis, Van Vliet & Van Mier (1994). Basically, we compute the principal stresses in a beam and define fracture using Mohr's circles. The law can be written in terms of beam forces as follows:

$$N_{crit} = \frac{1}{2} F_t \left[\frac{-1 \pm \sqrt{1 + 4(D/N_{act})^2}}{(D/N_{act})^2} \right] \tag{2}$$

In Figure 7 a graphical representation is given of the complete procedure. For the combination of the actual normal force (N_{act}) and shear force (D) computed in a beam element, the percentage of the critical combination (N_{crit}, D_{crit}) is computed. The beam having the highest percentage will be removed and the computation will be started again, in a similar way to our original approach.

As already mentioned, in most of the analyses carried out to date the fracture law Equation (1) was used. The simulations given in this paper are also based on this criterion. In section 6 it will be shown that under tensile and combined tensile/compressive states of stress the results of the simulations are independent of the choice of fracture law. Only for biaxial compressive states of stress is an important difference found. It will be obvious that at the level at which the model operates a single fracture law has to be selected. The second law, Equation (2), seems a good candidate, but some of the earlier work in which Equation (1) was used must be repeated before this choice can be made.

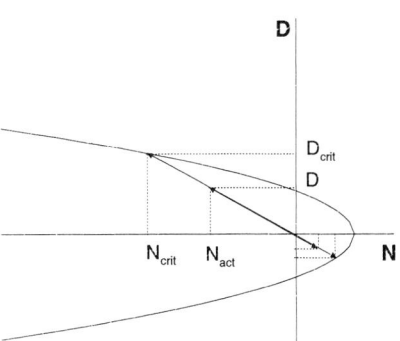

Fig. 7 Increase of failure load with factor N/N_{crit}, after Beranek & Hobbelman (1994).; after Van Vliet & Van Mier (1994).

3.3. Comparison of different approaches

We have studied the difference between different lattice types through the analysis of a single-edge-notched tensile specimen, see Figure 11a. We shall return to this geometry later on, when we attempt to explain the fracture mechanisms underlying softening of concrete and sandstone. The SEN specimen is 200 mm long, 100 mm wide and 50 mm thick. The specimen is loaded by giving the upper edge a uniform displacement. Thus the upper edge remains parallel to the lower edge during the entire analysis. Four different lattices were analysed, namely a regular triangular lattice with particle overlay, two different centre-particle lattices and a random lattice. The details of the material structure and the lattice size are as follows.

First of all we tried to keep the number of beam elements of each lattice roughly the same. The starting point was the number of elements necessary to construct the regular triangular lattice with particle overlay. The laboratory tests on SEN specimens were performed under displacement control using the average displacement of 4 LVDTs as feedback for the closed loop system (see Figure 11). The height of the lattice was kept equal to the measuring length (i.e. 35 mm); for the remainder of the specimen plane stress shell elements were used. In this way the computational effort can be reduced. The regular lattice was constructed using beams with a length of 5/3 mm. In order to acquire different beam properties, a particle overlay was generated containing aggregates between 0.25 mm and 8.0 mm, with a step size of 0.25 mm. Note that this particle structure was generated using the method described in Vervuurt et al. (1995), which was mentioned briefly in section 3.1.1. The same particle structure was used in the construction of the two centre-particle models and the regular lattice with particle overlay. Owing to the chosen beam length, all particles smaller than 3.0 mm were excluded. In this way a regular lattice consisting of 4269 beams was obtained. By taking a grid size $s = 5/3$ mm the number of beams for the random triangular lattice was 3854. As for the centre-particle lattice, it was not possible to make a mesh of about 4000 elements based on the generated grain structure. It was therefore decided to make two different centre-particle lattices. The coarse

Table 1 Number of beams removed for each lattice at peak load (I) and at beginning of horizontal part in descending branch (II).

Crack opening	Centre particle lattice (coarse)	Centre particle lattice (fine)	Random lattice	Regular lattice
I	25	25	50	200
II	75	100	100	275

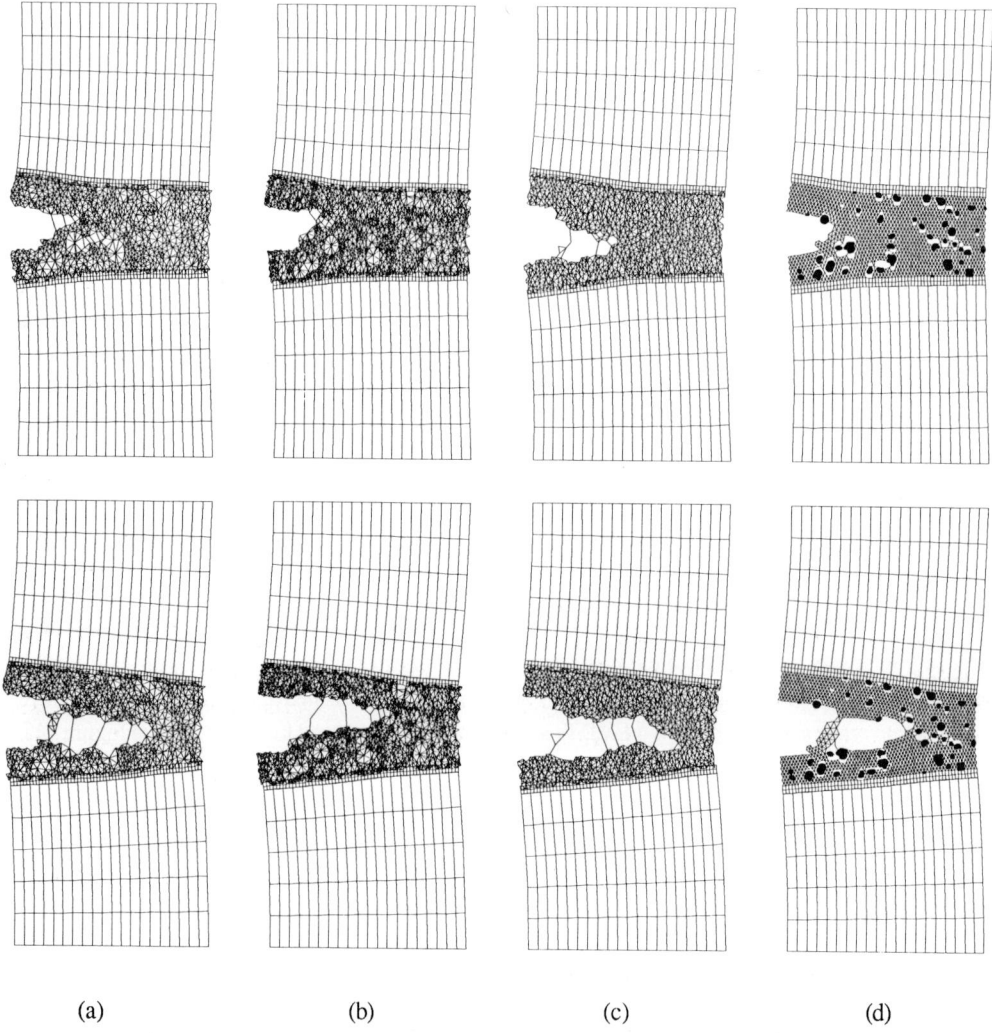

(a) (b) (c) (d)

Fig. 8 Analysis of crack growth in a single-edge-notcheed tensile specimen between parallel travelling end platens. Figure (a) shows crack growth at two stages for the fine centre particle model, (b) for the coarse centre particle model, (c) for the random triangular lattice without particle overlay, and (d) for the regular triangular lattice with particle overlay.

lattice, in which all particles smaller than 0.75 mm have been excluded, contains 3110 beams. On the other hand the fine centre-particle lattice was constructed including also the 0.5 mm particles. This gave a total number of 6968 beam elements. The cross-section of the beams of each lattice was determined in such a way that the Young's modulus of the un-notched lattices was equal to 30000 MPa. The results of the 4 simulations are shown in Figures 8 and 9. Figure 8 shows the deformed specimen at peak load (top figures) and at the beginning of the horizontal part of the descending branch (bottom figures) respectively. In Figure 9 the

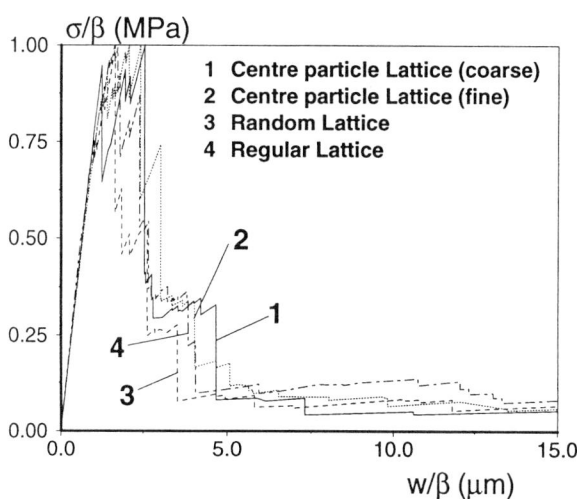

Fig. 9 Dimensionless stress-crack opening diagrams for the analysis of Figure 8

average stress-crack opening diagrams are plotted, the stresses and deformations being normalized using a factor β. β is the macroscopic peak strength found in an analysis. It should be mentioned that β depends on the type of lattice and therefore differs for each simulation. By normalizing both the stress and displacement using β, we find the same initial stiffness for all analyses and comparison of results is more straightforward.

Despite the rather large differences between the several types of lattices used, the shapes of the stress-crack opening diagrams do not differ very much (Figure 9). A closer look at the four curves shows a narrow peak regime for the random lattice and the coarse centre-particle lattice. This can also be concluded from Table 1 where the number of broken beams is given for the two stages mentioned above. The largest amount of broken beams corresponds to the regular lattice and is due to the microcracking found during the simulation (Figure 8d). The initial microcracks in the regular triangular lattice with particle overlay all appear in the bond zones, which have been assigned a low strength (see also section 4). In other words, the bond zones are the weakest link in the lattice. The amount of stresses that can be transferred in the tail of the stress-crack opening diagram is also related to the amount of detail in the finite element mesh. The fine meshes (regular lattice and fine centre-particle lattice) show a higher tail for the softening branch. It has to be mentioned that for each case only one simulation is performed. Since randomness is involved reliable results have to be obtained from more simulations. The behaviour, however, corresponds with the results of the impregnation experiments, which will be described in section 5.1.

4. Assessment of the Lattice Model Parameters

The advantage of the lattice model with different lattice geometries presented in some detail in the previous section is that all model parameters are *single-valued*. Therefore, the inverse process needed to determine the model parameters will become inherently more simple. The philosophy is that simplicity should prevail, especially in those cases where parameters only can be determined through an inverse modelling process. Note, however, the danger of inverse modelling: the parameter sets are in general not unique. Let us confine ourselves to the model with particle overlay, using Equation (1) as the fracture law. The parameters needed in the lattice model can be divided into two categories. The first category consists of those parameters related to the elastic stiffness of the lattice, the second set contains the parameters related to the fracturing of the lattice beams.

The main parameters in the first set are the cross-sectional area of the lattice beams and the Young's moduli of the material that is represented by the beams. For the specific case of a regular or random triangular lattice overlaid with the particle structure of the material, different Young's moduli must be specified for lattice beams appearing in each of the phases of the material, i.e. aggregate, matrix and bond zone (see also the previous section). Moreover, the beam thickness and depth must be chosen such that the Poisson's ratio and Young's modulus of the complete lattice match experimental observations. For a random triangular lattice an empirical relation was determined between the beam height-to-length ratio (h/ℓ) and the Poisson's ratio v for a random lattice of 50 * 50 nodes and a grid size of 5 mm; see Schlangen & Van Mier (1994). The following empirical formula was obtained:

$$v = \frac{4}{3+(h/\ell)^{\sqrt{3}}} - 1 \qquad (3)$$

For different lattices a similar procedure must be followed. For a regular triangular and isotropic random lattice exact solutions can be derived. For a regular square lattice $v = 0$.

The procedure is now as follows. As mentioned before, the length ℓ of the lattice beams is related to the smallest particle in the material structure. In general it suffices to take a beam length smaller than one third of the diameter of the smallest aggregate particle, Schlangen & Van Mier (1992b). In such a case the projected material structure determines the answer, and not the size of the lattice elements. The height of the beam can now be set through the previously determined relation between the h/ℓ-ratio and the Poisson's ratio. Subsequently the various stiffnesses are assigned to the lattice beams appearing in the three phases of the material, viz. aggregate, matrix and bond beams. We start by prescribing the correct ratios E_a/E_m and E_m/E_b corresponding to macroscopic values of the Young's moduli of the various phases. Thus, for example, for normal concrete we set $E_a/E_m = 70/25$ and for convenience we set the Young's modulus of the bond zone equal to the modulus of the matrix, thus $E_m/E_b = 25/25$. It should be mentioned that this is an approximation. The bond zone is generally more porous (at least for normal-weight concrete containing dense low-porosity natural aggregates), and the Young's modulus of the bond zone should be lower than the matrix modulus. At present we believe that this refinement of the model is not realistic in view of other uncertainties. Another point is that in reality a stochastic distribution of stiffness of the bond zone should be modelled. The same is true for the aggregate and matrix phases. Again this is a refinement that can be made in later stages. Note that at this stage the ratios of the various Young's moduli are important. The absolute elastic stiffness of a complete lattice is finally set by selecting an appropriate beam thickness b. It should be mentioned that the above procedure is the ideal procedure. In some of the analyses presented here this procedure was not fully followed, and deviating values of Poisson's ratio were obtained. However, in such cases we were mostly interested in fracture mechanisms and not so much in quantifying fracture.

So far only the beam size and stiffness have been adjusted to match the overall Poisson's ratio and Young's modulus of a complete lattice to the initial elastic stiffness of the material under consideration (in this case concrete). The second group of parameters consists of those related to the fracture law. The fracture law selected, Equation (1), is based on an effective stress determined from normal and flexural stresses in the beams. Essentially a beam is broken when the effective stress exceeds the tensile strength assigned to that particular beam. Thus the information needed comprises the fracture strengths of the beams appearing in the matrix, aggregate and bond zones of the material structure. Moreover, the parameter α in the fracture law should be determined. The procedure of assigning strength values to the beams is similar to the procedure followed to assign stiffnesses. First the strength ratios are assigned, corresponding to the realistic macroscopic strength of matrix, bond and aggregate. Thus not the absolute values are important, but rather the ratio's $f_{t,m} / f_{t,b}$ and $f_{t,a} / f_{t,b}$, where the subscript t stands for 'tensile' strength, and m, b and a refer to 'matrix', 'bond' and 'aggregate' respectively. For example, for a normalweight gravel concrete $f_{t,m} / f_{t,b} = 4$ and $f_{t,a} / f_{t,b} = 8$, derived from macroscopic strengths of 10, 5 and 1.25 MPa, are considered realistic.

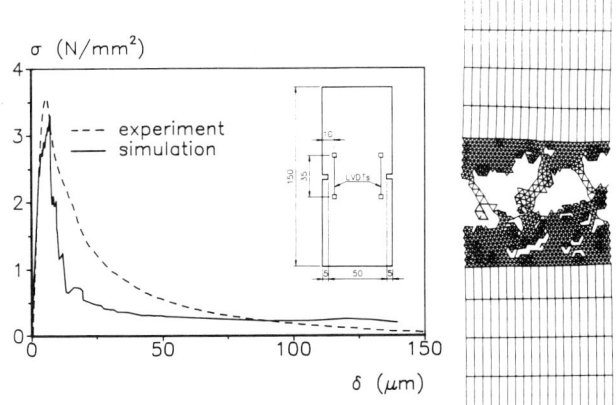

Fig. 10 Calibration of the model to an uniaxial tensile test on a concrete prism, after Schlangen & Van Mier (1992b).

The last and perhaps most difficult parameter is α, which regulates how much bending is taken into account. By changing α the failure mode of a beam can be changed from axial failure to predominantly flexural failure. From a number of analyses we have found that α hardly affects tensile fracture, Schlangen & Van Mier (1992b), but that it becomes more important in compressive fracture, Margoldová & Van Mier (1994) and Schlangen & Van Mier (1994). Increasing α from 0.1 to 1.5 has a profound effect on the macroscopic compressive fracture strength as compared with the tensile strength of the same random lattice. Moreover, the compressive behaviour changes from brittle to ductile when α decreases. For a regular lattice with particle overlay, Margoldová & Van Mier (1994) showed that the fracture mode in compression changed from vertical splitting to inclined cracking. These aspects are important for the final choice of the fracture law.

A uniaxial tension test on a small prismatic specimen (size 50*60*150 mm, with two 5 mm deep and 5 mm wide notches at half height) loaded between non-rotating end platens is used as calibration test to tune the macroscopic tensile strength of the lattice under consideration. Note that this calibration has to be repeated when different lattices are used, for example a regular lattice with a different beam length or a random lattice based on a different grid size etc. In Figure 10 the match between a numerical simulation and an experiment is shown for a regular triangular lattice with beam length $\ell = 5/3$ mm and particle overlay according to a Fuller distribution, where only the particles

Table 2 Overview of the parameters in the lattice model

Group 1	Parameters related to the elastic properties of the lattice		
Beam size	Length ℓ		related to the smallest feature (D_{min}) in the material structure
	Height h		related to the overall Poisson's ratio of the lattice (for example random lattice with $s = 5$ mm, Eq. (3))
	Thickness b		related to the overall Young's modulus of the lattice
Elastic moduli of the lattice beams	E_a E_m E_b		$E_m = E_b$ (simplification) ratio E_a / E_m is set according to realistic macroscopic data from uniaxial tension tests on aggregate and matrix materials
Group 2	Parameters related to the fracture properties of the lattice		
Fracture strength of the lattice beams	$f_{t,a}$ $f_{t,m}$ $f_{t,b}$		ratio's $f_{t,a}/f_{t,b}$ and $f_{t,m}/f_{t,b}$ are set according to realistic macroscopic values from laboratory scale uniaxial tensile tests
Flexural coefficient	α		tuning to standard tensile test (Schlangen & Van Mier (1992b))

between 3 and 8 mm are included. Note that in this calibration analysis, the model behaves still too brittle, but this can be largely attributed to excluding the small particles in the mesh and neglecting the third dimension, Van Mier et al. (1993) and Schlangen & Van Mier (1994).

In the ideal situation the same set of parameters as defined so far should also be used in analyses on different geometries and boundary conditions. In many publications we have shown that the model is capable of simulating the fracture mechanism in concrete and sandstone under tensile and combined tensile/shear loading, for example Schlangen & Van Mier (1992b, 1994), Van Mier et al. (1994a), Vervuurt & Van Mier (1994a), Nooru-Mohamed et al. (1994) and Vervuurt et al. (1994). In this paper we will include only the example of crack-face bridging in tension. For the particle overlay lattice with failure criterion (Equation 1) the complete set of parameters is given in Table 2.

One last remark should be made about the model parameters. It will be obvious that the fracture law parameters cannot be measured directly in any experiment but have to be retrieved through an inverse modelling technique. A new microscopy experiment is at present set up, in which we are trying to map the experimentally observed fracture patterns on computed fracture patterns, Vervuurt & Van Mier (1994b). Through a critical comparison of computed and measured crack patterns we hope to derive the properties of the interface between cement and different types of aggregates. These experiments will be outlined in another contribution to ICM7, Vervuurt & Van Mier (1995).

5. Crack Growth in Uniaxial Tension

Using the fracture law Equation (1) and the parameters defined in the previous section, we have analysed the crack growth process in single-edge-notched tensile specimens of similar geometry to that used before in the comparison of the different lattice types. The same specimen was used earlier in vacuum impregnation experiments, where we tried to elucidate the physical mechanisms underlying softening, Van Mier (1991a,b). Some results of these analyses are given below, including a discussion on the physical mechanisms and exaggerated brittleness of the model. Finally, the effect of boundary rotation stiffness on the crack growth process will be shown.

5.1. Crack growth analysis in a SEN tensile specimen

As already stated, using the model with fracture law Equation (1) and the particle overlay we analysed crack growth in the specimen of Figure 11a. The lattice is shown in Figure 11b. As in the previous analyses only the area where cracks were expected to grow was modelled as a lattice. The remainder of the specimen was modelled using plane stress elements available in the Finite Element Package DIANA that was used to solve the equations. The lattice was a triangular lattice of beam length 1.25 mm, whereas only aggregates between 2 and 8 mm were included in the mesh. The results have been shown before, Schlangen & Van Mier (1992a), but will be shown again here in a wider context. In Figure 12 the crack patterns computed at four different axial crack openings are shown. The four subsequent stages are indicated by the letters a-d in the stress crack opening diagram of Figure 13. Crack growth proceeds from distributed debonding near the aggregate particles, through growth of macro crack branches in stages b and c, to crack-face bridging in stage d. At this stage the specimen is fully cracked, but still a few connections exist,

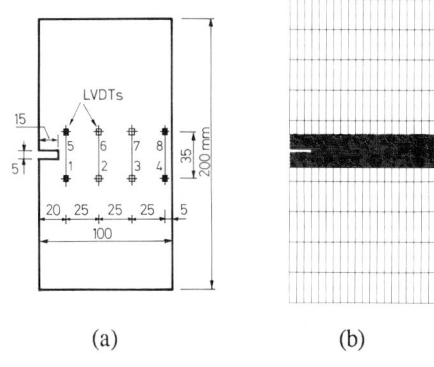

Fig. 11 SEN tensile specimen used in the impregnation experiments (a) and the element mesh (b). The numbers on Figure (a) refer to the locations of the LVDTs that were attached to the specimens surface. Test control was over the average of LVDTs 1,4,5 and 8; after Schlangen & Van Mier (1992a).

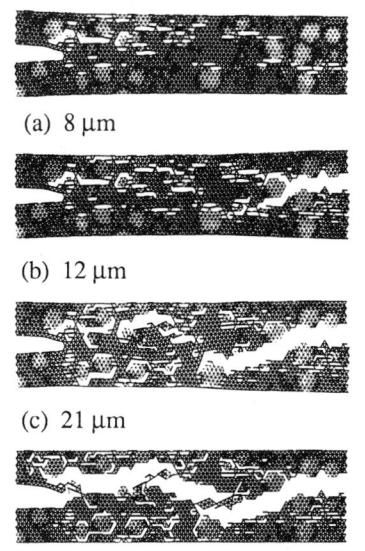

Fig. 12 Computed crack patterns at four stages of axial crack opening; after Schlangen & Van Mier (1992a).

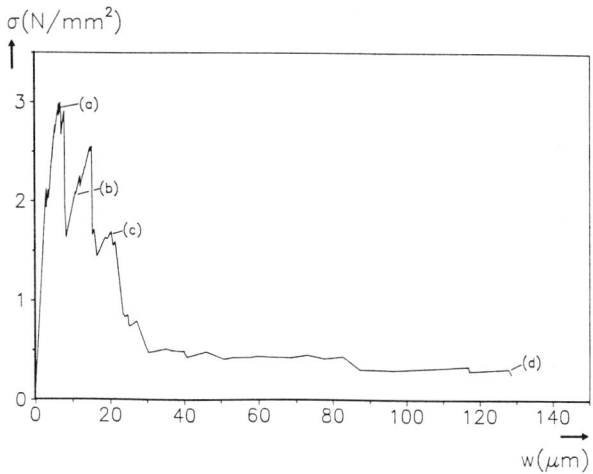

Fig. 13 Stress-crack opening diagram for the analysis of Figure 12. The four stages of crack growth plotted in the previous Figure are indicated by the letters a-d; after Schlangen & Van Mier (1992a).

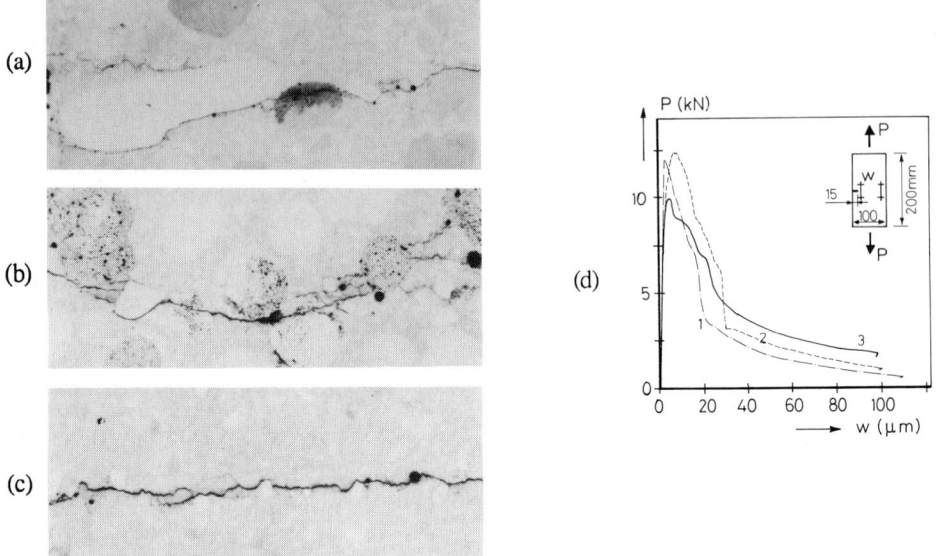

Fig. 14 Crack face bridging at w = 100 µm in 16 mm normal concrete (a), 12 mm lytag concrete (b) and 2 mm mortar (c). The load-displacement diagrams are shown in Figure (d); after Van Mier (1991b)..

which we refer to as crack-face bridges. As shown earlier in this paper, the type and amount of crack-face bridges depends on the type of lattice model used. So far the most reliable results (at least to our opinion) were obtained with the regular triangular lattice with particle overlay. Different results are found when the fracture properties of the matrix, aggregate and bond zones are changed. This was worked out in more detail in Schlangen & Van Mier (1992a, c).

Using the same specimen geometry, vacuum impregnation experiments were carried out on four different concretes. The results of these experiments were later complemented by a number of optical microscopy experiments using a QUESTAR QM-100 Remote Measurement System in combination with a fully automatic staging and imaging technique, see Vervuurt & Van Mier (1994b). Figure 14 shows some results of the impregnation experiments on three concretes with different maximum aggregate size. In Figure 14a-c internal crack patterns are shown in 2 mm mortar lytag lightweight concrete containing 12 mm lytag particles and sand with grain size up to 4 mm, and 16 mm normal concrete respectively. These crack patterns were all obtained by impregnating cracked specimens with fluorescenting epoxy at an average crack opening of 100 µm. The technique is fully described in Van Mier (1991a). In Figure 14d the three stress crack opening diagrams are shown for these three experiments. At the end of the softening branch the specimens were impregnated, cut open and the photographs of Figures a-c were taken under ultraviolet light. At a crack opening of 100 µm the specimen is fully cracked, but stress transfer is still possible. An important conclusion from the impregnation experiments was that the stress transfer in the tail of the diagram increases with increasing size of the stiff particles in the concrete, which is confirmed by many others, e.g. Petersson (1981). The mechanism responsible for the stress transfer in the tail of the softening diagram is visible in the three photographs: crack overlaps exist. These crack overlaps seem to develop preferentially near stiff aggregate particles. Note that in the lightweight concrete the main crack traverses the highly porous lightweight particles (which appear as speckled areas in the photograph). Bridges now appear around the stiff sand particles. This type of behaviour can also be simulated with the lattice model, but then of course the material structure overlay should be adjusted correspondingly. The crack-face bridges seem to be an important stress transfer mechanism in the tail of the softening diagram. Often it is argued that development

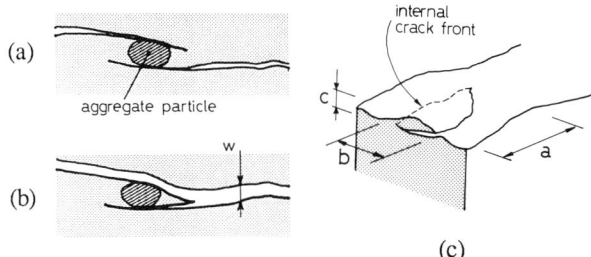

Fig. 15 *Crack overlap near a stiff aggregate particle (a), and failure mechanism (b). In reality the bridges are three-dimensional flaps which can be identified on a crack face after complete rupture (c); after Van Mier (1991a,b)..*

of a zone of distributed microcracks in front of the macrocrack tip is the main toughening mechanism in concrete. For example this mechanism was proposed, amongst others, in the seventies and early eighties by Hillerborg et al. (1976) and Bažant & Oh (1983). However, experimental verification has never been possible; see, for example, Mindess (1991). The bridging mechanism seems to make more sense, although it should be mentioned here that microcracking must occur. The impregnation experiments showed indirectly that microcrack weakening must occur at early stages of the fracture process; see Van Mier (1992). Steinbrech et al. (1991) claimed that they measured diffuse microcracking in ceramics. This issue remains open to debate, however, since much depends on the definition of a microcrack.

The crack-face bridging mechanism is shown schematically in Figure 15a in two dimensions. In Figure 15b a hypothesized 3D mechanism is shown. Basically a crack overlap will fail when one of the crack branches extends and joins the wake of the other crack, as shown in Figure 15a. Failure of a crack overlap was followed in a number of experiments using the QUESTAR QM100 remote

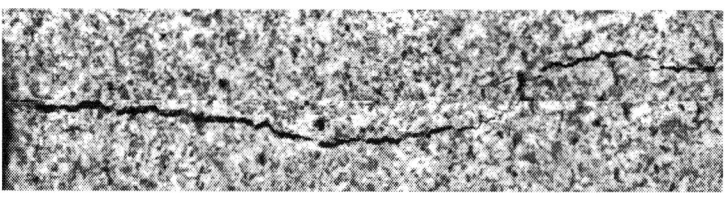

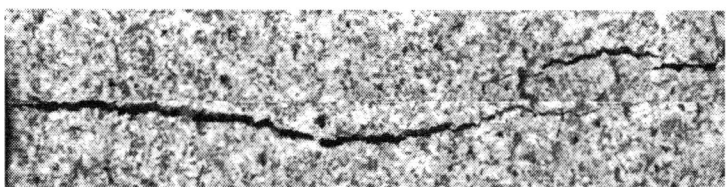

Fig. 16 *Example of crack face bridging in Colton sandstone. The images were taken from a uniaxial tension test, using our automatic scanning remote measurement system. Crack growth is shown at two stages (w = 120 µm and 150 µm), indicating clearly the existence of crack face bridges.*

measurement system; see Van Mier (1991b). These experiments confirmed the mechanism shown in Figure 15a.

The type of crack overlaps shown here for concrete have been found in a number of materials over a wide range of sizes. The crack geometry was earlier recognised by Sempere & MacDonald (1986). The relation with softening was first made in the experiments described above. In ceramics the same crack geometries can be found; see, for example, Swanson et al. (1987) and Steinbrech et al. (1991). In sandstone experiments Vervuurt found several good examples of crack-face bridging using our long-distance microscope, as shown in Figure 16. In general the overlap size ranges from a few micrometers in ceramics to some millimetres in concrete and rock. However, similar crack geometries can be found in the Mid-Atlantic ridge, and, for example, in the East African rift zone, but then at the kilometer scale. These latter mechanisms were addressed in the paper by Sempere & MacDonald (1986). A full overview is given in Van Mier (1991b, 1992).

In the laboratory experiments, interactions of the crack zone occur with the boundaries of the specimen. Recently we found that the amount of bridging depends on the freedom of the specimen ends to rotate in a tensile experiment. The above impregnation experiments were all carried out between parallel translating platens (referred to in the remainder of this paper as 'fixed' platens). Before showing some of these results, the mechanisms which cause the exaggerated brittleness of the model will be clarified. This can also be seen from a comparison of the stress-crack opening diagrams of the impregnation tests (Figure 14d) with the simulated diagrams shown in Figures 9 and 13.

5.2. Brittleness in the model

There are at least two important reasons for the exaggerated brittleness of the lattice model. Both reasons are strongly related to the computational effort needed for a simulation. The first reason is that not all details of the material structure have been taken into account. Most important in the previous analyses was that a lower cut-off was made at 2 mm, i.e. all aggregates smaller than 2 mm were omitted from the analyses. This is important, since the length of the beams in the lattice and therefore the number of beam elements is directly related to the size of the smallest aggregate particle included in the material structure. This occurs for any lattice that we want to use, as is clear also from the comparative analyses of Figure 8. From this figure it is obvious, for example, that in the centre-particle lattice the number of beams increases massively when a lower cut-off is made for the particle size. However, the amount of detail in the crack patterns increases substantially (although this is more visible in the particle overlay method). The effect of including smaller particles in the mesh was demonstrated in a qualitative analysis by Schlangen & Van Mier (1993) and later for

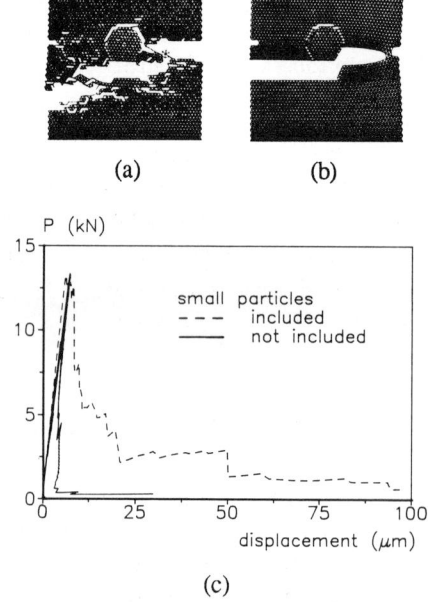

Fig. 17 Effect of small particles in the model. Mesh with (a) and without (b) small particles, and (c) computed load-displacement diagrams; after Schlangen & Van Mier (1993)..

the standard test geometry of Figure 10 by Van Mier et al. (1993). Figure 17 shows the result of the qualitative analysis. A lattice (80 x 80 mm) containing a single large aggregate (d = 20 mm) was subjected to uniaxial tension. This analysis was compared with a simulation where the large particle was surrounded by numerous randomly placed 4 mm particles. The two meshes are shown in Figures 17a and b. The computed load-displacement diagrams are shown in Figure 17c. The comparison shows that the amount of detail in the crack patterns increases massively when the small particles are added. Moreover, the brittleness decreases, i.e the slope of the falling branch decreases, as can be seen from Figure 17c. In other words, the brittleness would disappear to some extent in the previous analyses of the SEN specimen if the small details of the material structure were included in the analysis. It will be evident that the computational effort will increase substantially when all small details in the material structure are included.

The second effect is what is referred to as 3D effect; see Schlangen & Van Mier (1994). The crack growth process is in reality a three-dimensional process. This was clearly demonstrated by the impregnation experiments; see Van Mier (1990). In a recent three-dimensional analysis it was found that the crack front spreads non-uniformly through the specimen cross-section, much in line with earlier observations. The brittleness in the analysis was not reduced however, as a relatively coarse lattice was used. Again this was done to limit the computational effort. A full 3D comparison will only become possible in the future, when more powerful computers and algorithms become available.

The exaggerated brittleness may perhaps also be attributed to the typical fracture law used in the simulations. Future work should indicate whether a change of fracture law, as suggested for example in section 3.2 and partly elaborated in section 6, would also lead to less severe brittleness.

5.3. Effect of boundary rotations

Using 'fixed' loading platens in a tensile experiment has an important effect on the crack growth process. Owing to the heterogeneity of the material, crack growth will always start from one side, even in a completely symmetrical test set-up, Van Mier (1986). This has become an important research issue. Many have followed and tried to analyse non-uniform opening in the 'fixed' tensile tests, for example Hordijk et al. (1987), Rots & De Borst (1989), Rossi (1989) and Bažant & Cedolin (1993). Recently an experiment was proposed by Carpinteri & Ferro (1993), in which a dog-bone shaped specimen was subjected to uniaxial tension. A complicated loading system consisting of three controlled hydraulic actuators was used to avoid non-uniformities in the fracture zone. A central actuator was used for loading the specimen; the other two actuators were used to correct non-uniformities that might occur in-plane and out-of-plane. According to Carpinteri & Ferro, in this test all non-uniformities have been removed and an ideal softening diagram is retrieved. We analysed the experiment by Carpinteri & Ferro with our lattice model, see Vervuurt & Van Mier (1994a). Owing to computational limitations we could only perform a 2D analysis within a reasonable amount of time. In addition to the boundary condition used by Carpinteri and Ferro, we analysed three other conditions. These are summarized in Figure 18. In Figures 18a and b the deformation in a central zone of 35 mm and 50 mm length respectively is kept constant. Figure 18b corresponds to the loading condition in the experiment by Carpinteri and Ferro. We compared these cases to the 'fixed' condition, where the specimen ends are kept parallel to one another (Figure 18c), and to the freely rotating specimen which is shown in Figure 18d. The results have been exhaustively published in Van

Mier et al. (1994a,b) and Vervuurt & Van Mier (1994a). Here we show the effect on fracture energy; see Table 3. The fracture energy is defined as the area under the stress-crack opening diagram, where crack opening is measured over a 35 mm gauge length. The fracture energies were determined up to a crack opening of 100 µm. For each boundary condition two analyses were made, using two different particle structures. The results of all eight analyses are included in Table 3. The results clearly indicate an effect of rotational stiffness on fracture energy. The fracture energy decreases when the rotational freedom increases. Note that in the analyses A, B and C the flexural stiffness of the part of the specimen in the control length determines the rotational freedom of the crack zone.

Recently we carried out a number of experiments between 'fixed' and freely rotating loading platens. In these experiments we found that the fracture energy under freely rotating loading platens decreases by about 30 to 40 % as compared with specimens that are loaded between 'fixed' end platens. A key figure is reproduced in Figure 19, where the evolution of fracture energy with average crack opening is shown. The diagram for fixed boundaries is higher than the curve for rotating boundaries. The results are for 8 mm concrete, but similar observations were made for Colton

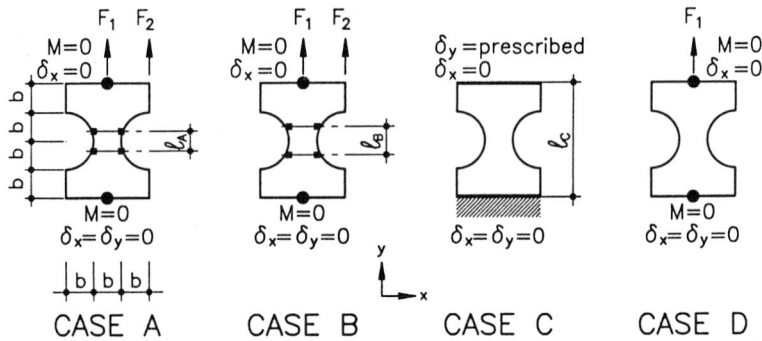

Fig. 18 Specimen geometry tested by Carpinteri & Ferro (1993), and the four different boundary conditions analysed with the lattice model; after Van Mier et al. (1994a).

Table 3 Computed fracture energies.
    The deformations are based on a measuring length of 35 mm in all cases. The fracture energy has been calculated as the area under the stress-crack opening diagram up to 100 µm. The cross-sectional area is taken equal to 50*100 = 5000 mm^2.

Simulation	grain-structure	G_f (N/m)
CASE A (ℓ_{meas} = 35 mm)	1	40.0
	2	44.0
CASE B (ℓ_{meas} = 50 mm)	1	40.6
	2	43.8
CASE C (fixed)	1	35.3
	2	32.5
CASE D (rotating)	1	29.8
	2	11.6

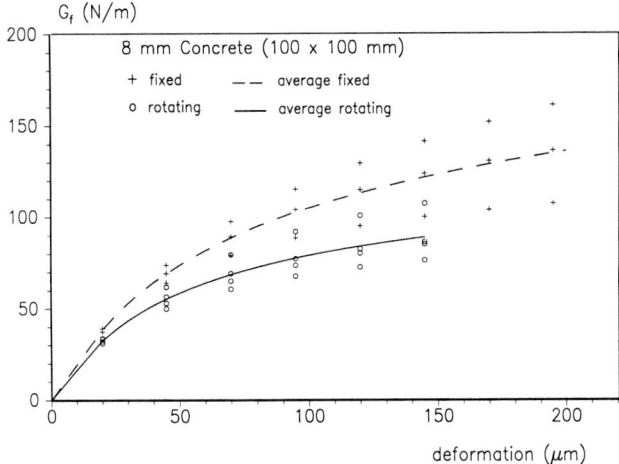

Fig. 19 Fracture energy versus axial crack opening for tests between 'fixed' and freely rotating loading platens. The specimens were 100 mm diameter and 100 mm long cylinders with a 5 mm deep circumferential notch at half height; after Van Mier et al. (1994b).

sandstone and Felser sandstone; see Van Mier et al. (1994b). In addition to the effect on fracture energy, the rotational freedom of the loading platens also affects the post-peak stress-crack opening diagram. This is shown qualitatively in Figure 20. For the full results the reader is referred to Van Mier et al. (1994b). Basically, a bump is found in the softening branch when 'fixed' boundaries are used, whereas a smooth softening curve is measured under freely rotating conditions. These results confirm what was hypothesized in Van Mier (1986b) and later measured by Daerga (1992). Unfortunately the results obtained by Daerga were too limited to permit any quantitative conclusions, which led us to repeat the experiment. The numerical analyses mentioned above indicated that the crack growth processes also depend on the boundary conditions. Under rotating platens a single crack develops from one side of the specimen. Under 'fixed' boundaries, a crack also starts at one side of the specimen but is arrested by a bending moment. This bending moment is the result of the gradually increasing load eccentricity when the crack propagates. At some point in time the specimen must crack on the other side, and two interacting cracks develop. These interacting cracks were observed in many experiments; see, for example, Van Mier & Nooru-Mohamed (1990). Note that the analyses based on a softening smeared crack model by Rots & De Borst (1989) showed no difference in stress-crack opening behaviour when the boundary conditions were changed from fixed to freely rotating. On the other hand the analyses by Rossi (1989) showed the correct result.

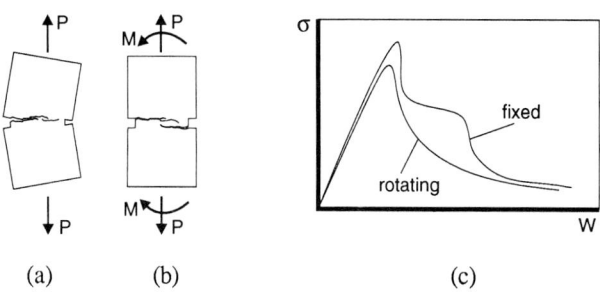

Fig. 20 Effect of boundary conditions on softening: (a) freely rotating boundaries, and (b) 'fixed' boundaries. In Figure (c) the stress-crack opening diagrams are shown.

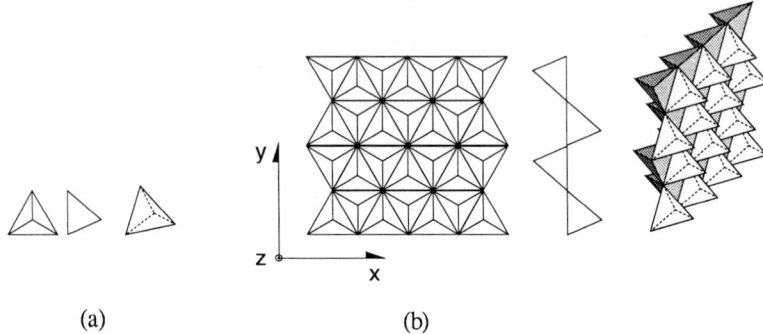

Fig. 21 Front view, side view and isometry for the 1x1 mesh (a) and the 4x4 mesh (b); after Van Vliet & Van Mier (1994)

6. Simulating Out-of-plane Fracture under Global Compression

So far all analyses have concerned tensile fracture. Only the fracture law Equation (1) was used, and the results seem quite promising. However, when we started to examine compressive fracture of concrete, it was found that the load-displacement response, the peak load and the post-peak response depend to a large extent on the coefficient α in the fracture law; see Schlangen & Van Mier (1994) and Margoldová & Van Mier (1994). This has more or less forced us to check whether the fracture law that we have used is the correct one for generalized stress. Inspired by Beranek & Hobbelman (1994), we decided to carry out a number of analyses with the fracture law Equation 2. As a test we tried to determine the shape of the biaxial failure contour for the lattices of Figure 21. In order to simulate out-of-plane fracture in biaxial compression, a three-dimensional lattice was adopted. As a basic element a tetrahedron was selected. Three different lattice sizes were analysed, namely a mesh consisting of a single tetrahedron, a 4x4 mesh containing 28 tetrahedrons, and a 21x24 mesh containing 984 tetrahedrons. In addition to the fracture law we investigated the effect of the rotational freedom of the nodes in the lattice. Full details of all these analyses are given in Van Vliet & Van Mier (1994); here we will show only the effect of changing the fracture law. In Figure 22 the computed biaxial failure contours are shown for three different sets of analyses. The shear force criterion (Equation 2) is compared with the moment criterion, with $\alpha = 0.5$ and 0.005 respectively. In this example a 4x4 mesh was used; the rotations of all nodes in the plane $z = 0$ were suppressed. The result is quite straightforward. When the shear force criterion is used, it is found that the biaxial failure locus is closed in the biaxial compressive regime, and out-of-plane fracture can be simulated. However, using the moment criterion (Equation 1), the biaxial failure contour is open in the biaxial

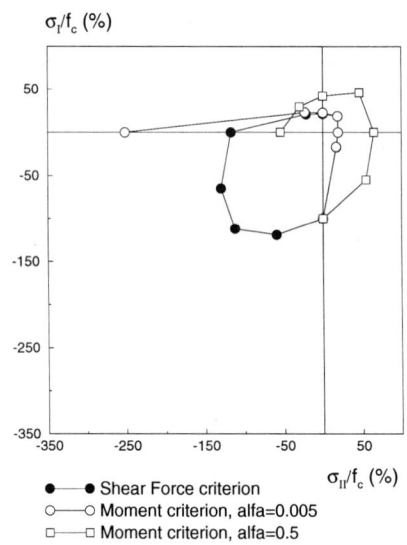

Fig. 22 Effect of choice of fracture law on the biaxial failure contour. A 4x4 mesh of tetrahedrons was used in these analyses; after Van Vliet & Van Mier (1994).

compression regime. These results suggest that a change of fracture law is necessary in order to describe fracture under generalized stress. Note that the difference between the two fracture laws is minimal in the biaxial tensile and tension-compression regimes. Only for high values of α does the shape of the failure contour become distorted in the tensile regimes. This would suggest that the tensile results presented in the first part of this paper would remain the same when the shear fracture criterion is used, but this must still be verified. The behaviour of the lattice in compression is at present being studied in more detail. Especially edge effects in the lattice and the influence of the elements in the third dimension need some further investigation. It is mentioned that a single fracture law must suffice to describe the behaviour of the material at the particle level. Here it is only suggested that there is reason to change the earlier normal load bending moment criterion according to Equation (1).

7. Conclusion

This paper gives an overview of our recent work on fracture of concrete and rock. Both experiments and analyses with a recently developed lattice model are used to obtain a more fundamental insight into the fracture process. The lattice model is an excellent tool for studying fracture in laboratory-scale specimens. It is essential is that the exact boundary conditions in the experiment simulated are taken into account in the numerical analyses. The numerical and experimental results obtained to date suggest that fracture in brittle disordered materials such as concrete and rock is a highly complex, non-uniform process. After distributed microcracking, macrocrack branches develop. The macrocrack branches do not coalesce, but crack overlaps develop. These crack overlaps develop primarily near stiff aggregate particles in the concrete. This type of crack-face bridging seems to be the main toughening mechanism in concrete and rock and causes the long stable tail in the softening diagram. Moreover, the same has also been observed in some ceramics as well. Owing to the non-uniformity of the fracture process, significant boundary and specimen size effects are observed. In the paper the effect of rotational freedom of the specimen boundaries in uniaxial tensile tests is examined. It is found that the crack density increases when the specimen is loaded between parallel translating loading platens ('fixed' boundaries). The resulting difference in fracture energy can be as high as 30 to 40 %.

The lattice model has given excellent results in tension. Noteworthy is the difference between the centre-particle lattice and the lattice with particle overlay. In the centre-particle lattice, the analysis is based on ideas concerning load transfer and interaction between rigid particles in concrete (see Figure 1). In the particle overlay method we identified stiff aggregates and weak interface zones directly in the model. This latter approach seems to give more realistic crack mechanisms, while the stress-crack opening behaviour remains almost identical. In order to increase the range of applicability of the lattice model, a new fracture law has been implemented. The effect of this fracture law on the stress-displacement behaviour has not been studied to date but will be subject to further study.

Acknowledgement

The financial support received for parts of this work from the Dutch Technology Foundation (STW) is gratefully acknowledged. The experiments were carried out by A. Elgersma and G. Timmers; their expert assistance is highly appreciated.

References

Bažant, Z.P. and Oh, B.-H. (1983), Crack band theory for fracture of concrete, *Mater. Struct. (RILEM)*, **16**, 155.

Bažant, Z.P. and Cedolin, L. (1993), Why direct tension test specimens break flexing to the side, *J. Struct. Eng. (ASCE)*, **119**, 1101.

Bentz, D.P., Garboczi, J. and Stutzman, P.E. (1992), Computer modelling of the interfacial zone in concrete, in *Interfaces in Cementitious Composites*, Maso, J.C., Ed., E&FN Spon/ Chapman & Hall, London/New York, 107.

Beranek, W.J. and Hobbelman, G.J. (1994), Constitutive modelling of structural concrete as an assemblage of spheres, in *Computer Modelling of Concrete Structures*, Mang, H., Bi_ani_, N. and De Borst, R., Eds., Pineridge Press, Swansea, 37.

Berg, A. and Svensson, U. (1991), Datorsimulering och analysis av brottförloppi en heterogen materialstruktur (Simulation of fracture in heterogeneous materials), *Report no. TVSM-5050*, Lund Institute of Technology, Department of structural Engineering, Lund, Sweden (in Swedish).

Burt, N.J. and Dougill, J.W. (1977), Progressive failure in a model heterogeneous medium, *J. Eng. Mech. Div. (ASCE)*, **103**, 365.

Carpinteri, A. and Ferro, G. (1993), Apparent tensile strength and fictitious fracture energy of concrete: A fractal geometry approach to related size effects, in *Fracture and Damage of Concrete and Rock (FDCR-2)*, Rossmanith, H.P., Ed., Chapman & Hall/E&FN Spon, London/ New York, 86.

Charmet, J.C., Roux, S. and Guyon, E., Eds. (1990), *Disorder and Fracture*, Plenum Press, New York.

Daerga, P.A. (1992), Some experimental fracture mechanics studies in mode I of concrete and wood, *Licentiate thesis*, Luleå University of Technology, Sweden.

Gerstle, K.H., Linse, D.L., Bertacchi, P., Kotsovos, M.D., Ko, H.-Y., Newman, J.B., Rossi, P., Schickert, G., Taylor, M.A., Traina, L.A., Zimmerman, R.M. and Bellotti, R. (1978), Strength of concrete under multiaxial stress states, in *Proceedings Douglas McHenry Int'l. Symposiun on 'Concrete and Concrete Structures'*, ACI SP-55, American Concrete Institute, Detroit, 103.

Hansen, A., Roux, S. and Herrmann, H.J. (1989), Rupture of central force lattices, *J. Phys. France*, **50**, 733.

Herrmann, H.J. (1991), Patterns and scaling in fracture, in *Fracture Processes in Concrete, Rock and Ceramics*, Van Mier, J.G.M., Rots, J.G. and Bakker, A., Eds., Chapman & Hall/E&FN Spon, London/ New York, 195.

Hillerborg, A., Modéer, M. and Petersson, P.-E. (1976), Analysis of crack formation and crack growth in concrete by means of fracture mechanics and finite elements, *Cem. & Conc. Res.*, **6**, 773.

Hordijk, D.A., Reinhardt, H.W. and Cornelissen, H.A.W. (1987), Fracture mechanics parameters of concrete from uniaxial tensile tests as influenced by specimen length, in *Pre-Proceedings SEM/RILEM Conference on 'Fracture of Concrete and Rock'*, Shah, S.P. and Swartz, S.E., Eds., Society of Experimental Mechanics, Bethel (CT), USA, 138.

Horii, H. and Nemat-Nasser, S. (1985), Compression-induced microcrack growth in brittle solids: axial splitting and shear failure, *J. Geophys. Res.*, **90**, 3105.

Hrennikoff, A. (1941), Solution of problems of elasticity by the framework method, *J. Appl. Mech.*, A169.

Hughes, B.P. and Ash, J.E. (1970), Anisotropy and failure criteria for concrete, *Mater. Struct. (RILEM)*, **3**, 371.

Kemeny, J.M. and Cook, N.G.W. (1991), Micromechanisms of deformation in rocks, in *Toughening Mechanisms in Quasi-Brittle Materials*, Shah, S.P., Ed., Kluwer Academic, Dordrecht, 155.

Kotsovos, M.D. (1983), Effect of testing techniques on the post-ultimate behaviour of concrete in compression, *Mater. Struct. (RILEM)*, **16**, 3.

Kupfer, H. (1973), Das Verhalten des Betons unter Mehrachsiger Kurzzeit-belastung unter Besonderer Berücksichtigung des Zweiachsiger Beanspruchung, *Deutscher Ausschuss für Stahlbeton*, Vol. 229, Berlin (in German).

Margoldová, J. and Van Mier, J.G.M. (1994), Simulation of compressive fracture in concrete, in *ECF10 'Structural Integrity: Experiments, Models, Applications'*, Schwalbe, K.-H. and Berger C., Eds., EMAS Publishers, Warley, UK, 1399.

Mindess, S. (1989), Interfaces in concrete, in *Materials Science of Concrete I*, Skalny, J.P., Ed., The American Ceramic Society Inc., Westerville (OH), 163.

Mindess, S. (1991), Fracture process zone detection, in *Fracture Mechanics Test Methods for Concrete*, Shah, S.P. and Carpinteri, A., Eds., Chapman & Hall, London, Chap. 5.

Nihei, K.T., Myer, L.R., Kemeny, J.M., Liu, Z. and Cook, N.G.W. (1994), Effects of heterogeneity and friction on the deformation and strength of rock, in *Fracture and Damage of Quasibrittle Structures*, Bažant, Z.P., Bittnar, Z., Jirásek, M. and Mazars, J., Eds., E&FN Spon, London/New York, 479.

Nooru-Mohamed, M.B., Schlangen, E. and Van Mier, J.G.M. (1993), Experimental and numerical study on the behavior of concrete subjected to biaxial tension and shear, *Adv. Cement Based Mater.*, **1**, 22.

Petersson, P.-E. (1981), Crack growth and development of fracture zones in plain concrete and similar materials, *Report TVBM-1006*, Lund Institute of Technology, Sweden.

Roelfstra, P.E. (1989), Simulation of strain localization processes with numerical concrete, in *Cracking and Damage*, Mazars, J. and Bažant, Z.P., Eds., Elsevier Applied Science, London/New York, 79.

Rossi, P. (1989), Numerical modelling of cracking using a non-deterministic approach, In *Fracture Toughness and Fracture Energy*, Mihashi, H., Takahashi, H. and Wittmann, F.H., Eds., Balkema, Rotterdam, 383.

Rots, J.G. and De Borst, R. (1989), Analysis of concrete fracture in 'direct' tension, *Int. J. Solids Struct.*, **25**, 1381.

Schlangen, E. (1995), Computational aspects of fracture simulations with lattice models, in *Proceedings FraMCoS II*, Wittmann, F.H., Ed., Aedificatio Publishers, in print.

Schlangen, E. and Van Mier, J.G.M. (1992a), Experimental and numerical analysis of micro-mechanisms of fracture of cement-based composites, *Cem. & Conc. Composites*, **14**, 105.

Schlangen, E. and Van Mier, J.G.M. (1992b), Micromechanical analysis of fracture of concrete, *Int. J. Damage Mech.*, **1**, 435.

Schlangen, E. and Van Mier, J.G.M. (1992c), Numerical study of the influence of interfacial properties on the mechanical behaviour of cement-based composites, in *Interfaces in Cementitious Composites*, Maso, J.C., Ed., Chapman & Hall/E&FN Spon, London/New York, 237.

Schlangen, E. and Van Mier, J.G.M. (1993), Lattice model for simulating fracture of concrete, in *Numerical Models in Fracture Mechanics of Concrete*, Wittmann, F.H., Ed., Balkema, Rotterdam, 1993.

Schlangen, E. and Van Mier, J.G.M. (1994), Fracture simulations in concrete and rock using a random lattice, in *Computer Methods and Advances in Geo-mechanics*, Siriwardane, H. and Zaman, M.M., Eds., Balkema, Rotterdam, 1641.

Sempere, J.-C. and Macdonald, K.C. (1986), Overlapping spreading centers: implications from crack growth simulation by the displacement discontinuity method, *Tectonics*, **5**, 151.

Steinbrech, R.W., Dickerson, R.M. and Kleist, G. (1991), Characterization of the fracture behaviour of ceramics through analysis of crack propagation studies, in *Toughening Mechanisms in Quasi Brittle Materials*, Shah, S.P., Ed., NATO ASI series, Vol. E-195, Kluwer Academic Publishers, Dordrecht, 287.

Stroeven, P. (1973), Some aspects of the micromechanics of concrete, *Ph.D. thesis*, Delft University of Technology, The Netherlands.

Swanson, P.L., Fairbanks, C.L., Lawn, B.R., Mai, Y.-W. and Hockey, B.J. (1987), Crack-interface grain bridging as a fracture resistance mechanism in ceramics: I, experimental study on alumina, *J. Am. Ceram. Soc.*, **70**, 279.

Van Mier, J.G.M. (1984), Strain-softening of concrete under multiaxial loading conditions, *Ph.D. thesis*, Eindhoven University of Technology, The Netherlands.

Van Mier, J.G.M. (1986a), Multiaxial strain-softening of concrete, part I: Fracture, part II: Load-histories, *Mater. Struct. (RILEM)*, **19**, 179.

Van Mier, J.G.M. (1986b) Fracture of Concrete under Complex Stress. *HERON*, **31**(3), 1.

Van Mier, J.G.M. (1990), Fracture process zone in concrete: A three dimensional growth process, in *ECF8 'Fracture Behaviour and Design of Materials and Structures'*, Firrao, D., Ed., EMAS Publishers, Warley, UK, 567.

Van Mier, J.G.M. (1991a), Mode I fracture of concrete: Discontinuous crack growth and crack interface grain bridging, *Cem. & Conc. Res.*, **21**, 1.

Van Mier, J.G.M. (1991b), Crack face bridging in normal, high stength and lytag concrete, in *Fracture Processes in Concrete, Rock and Ceramics*, Van Mier, J.G.M., Rots, J.G. and Bakker, A., Eds., Chapman & Hall/E&FN Spon, London/New York, 27.

Van Mier, J.G.M. (1992), Scaling in tensile and compressive fracture of concrete, in *Applications of Fracture Mechanics to Reinforced Concrete*, Carpinteri, A., Ed., Elsevier Applied Science, London/New York, Chap. 5.

Van Mier, J.G.M. and Nooru-Mohamed, M.B. (1990), Geometrical and structural aspects of concrete fracture, *Eng. Fract. Mech.*, **35**, 617.

Van Mier, J.G.M., Schlangen, E. and Vervuurt, A. (1993), Analysis of fracture mechanisms in particle composites, in *Micromechanics of Concrete and Cementitious Composites*, Huet, C., ed., Presses Polytechniques et Universitaires Romandes, Lausanne, 159.

Van Mier, J.G.M., Vervuurt, A. and Schlangen, E. (1994a), Crack growth simulations in concrete and rock, in *Probabilities and Materials. Tests, Models and Applications*, Breysse, D. Ed., NATO-ASI Series E Applied Sciences, Vol. 269, Kluwer Academic Publishers, Dordrecht, 377.

Van Mier, J.G.M., Vervuurt, A. and Schlangen, E. (1994b), Boundary and size effects in uniaxial tensile tests: a numerical and experimental study, in *Fracture and Damage of Quasibrittle Structures*, Bažant, Z.P., Bittnar, Z., Jirásek, M. and Mazars, J., Eds., E&FN Spon, London/New York, 289.

Van Vliet, M.R.A. and Van Mier, J.G.M. (1994), Comparison of lattice type fracture models for concrete under biaxial compression, in *Size-Scale Effects in the Failure Mechanisms of Materials and Structures*, Carpinteri, A., Ed., E&FN Spon, London/New York (in press).

Vervuurt, A. and Van Mier, J.G.M. (1994a), Experimental and numerical analysis of boundary effects in uniaxial tensile tests, in *Localized Damage III - Computer Aided Assessment and Control*, Aliabadi, M.H., Carpinteri, A., Kalisky, S. and Cartwright, D.J., Eds., Computational Mechanics Publications, Southampton, 3.

Vervuurt, A. and Van Mier, J.G.M. (1994b), An optical technique for surface crack measurements of composite materials, in *Recent Advances in Experimental Mechanics*, Silva Gomes, J.F., Branco, F.B., Martins de Brito, F., Gil Saraiva, J., Ludes Eusebio, M., Sousa Cirne, J. and Correia da Cruz, A., Eds., Balkema, Rotterdam, 437.

Vervuurt, A., Van Mier, J.G.M. and Schlangen, E. (1994), Analysis of anchor pull-out in concrete, *Mater. Struct. (RILEM)*, **27**, 251.

Vervuurt, A. and Van Mier, J.G.M. (1995), in *Proceedings ICM7*, The Hague, May 30 - June 2, 1995.

Vervuurt, A., Van Vliet, M.R.A., Van Mier, J.G.M. and Schlangen, E. (1995), Simulations of tensile fracture in concrete, in *Proceedings FraMCoS II*, Wittmann, F.H., Ed., Aedificatio Publishers, in print.

Vile, G.W.D. (1968), The strength of concrete under short term static biaxial stress, in *Proceedings Int'l. Conference on 'The Structure of Concrete'*, Cement & Concrete Association, London, 275.

Vonk, R.A. (1992), Softening of concrete loaded in compression, *Ph.D. thesis*, Eindhoven University of Technology, The Netherlands.

Vonk, R.A., Rutten, H.S., Van Mier, J.G.M. and Fijneman, H.J. (1989), Influence of boundary conditions on softening of concrete loaded in compression, in *Fracture of Concrete and Rock - Recent Developments*, Shah, S.P., Swartz, S.E. and Barr, B., Eds., Elsevier Applied Science, London/New York, 711.

Walraven, J.C. (1980), Aggregate interlock: A theoretical and experimental analysis, *Ph.D. thesis*, Delft University of Technology, The Netherlands.

Zubelewicz, A. and Bažant, Z.P. (1987), Interface element modeling of fracture in aggregate composites, *J. Eng. Mech. (ASCE)*, **113**, 1630.

*W. Michaeli**

Relations between the Mechanical Behaviour of Polymers and their Processing Methods and Conditions

Reference: Michaeli, W. (1995), Relations between the Mechanical Behaviour of Polymers and their Processing Methods and Conditions. In: *Mechanical Behaviour of Materials* (ed. A. Bakker), Delft University Press, Delft, The Netherlands, pp. 127-142.

Abstract: Processing significantly affects the mechanical properties of a polymer product. The properties are generally related to the orientation of an extruded film or the crystalline structure of an injection-moulded part. It is therefore of interest to examine in detail the effects of processing on the quality of the mouldings. The paper discusses and gives examples of two different approaches to the understanding of these relations. The first approach is an experimental method using physical processing data such as temperature, cooling rate, shear rate derived from the machine setting and the machine and mould geometries. These are correlated empirically with the part properties. The second approach is the theoretical prediction of mechanical part properties on the basis of computer modelling of the molecular orientation and the developing crystalline structure. This is an encouraging step towards a better understanding of the relations between processing and properties.

1. Introduction

It is very well known - and also theoretically understood - that by stretching a semi-crystalline polymer film at a temperature a little below the crystallization temperature the mechanical properties, e.g. Young's modulus, can significantly be improved. This is essential for example in the production of video tape film. It can also be seen that when cooled at different cooling rates (high or low) some materials may remain totally transparent or even become opaque due to the development of a different crystallinity. By mechanically testing these samples or products one can often also detect marked differences in mechanical properties. This is also of importance in polymer film production. Similarly when we analyse the mechanical performance of a complex-shaped technical part, e.g. an injection-moulded clutch-pedal made of glass-fiber-reinforced polyamide, the specialist at any rate knows that its mechanical behaviour is very much dependent on the specific gating situation and the selected injection processing conditions.

Perhaps the most fundamental and essential question in polymer processing is to know how to handle a specific material, how to run and how to adjust a process in order to achieve a desired product quality. New polymers, computer-integrated manufacturing and quality requirements everyday call for a fresh, up-to-the-minute answer to this question. The answer is very complex and requires fundamental knowledge about how, in the case of a given polymer, the structure of a product correlates with its processing conditions. As shown in the above examples, these crucially affect the final properties and product performance. When we talk about processing conditions we have to keep in mind that they are often dependent on the polymer itself, the specific design of the processing equipment, the quality of the control systems, the handling and pretreating, e.g. drying, compounding, evapozing, and the environmental conditions.

---
* *Institut für Kunststoffverarbeitung, RWTH Aachen, Pontstraße 49, 52062 Aachen, Germany*

While trying to understand processing better, we have to take care to recognize that changing one parameter during processing can also affect other final product properties, either positively or negatively. We therefore have to identify and study the individual influences for a given material. It is first necessary to optimize the design of a plastics part before seeking optimized material processing. Optimum part design is not, however, the objective of this paper. The final properties of a plastics part made from a given and defined polymer are independent of the production process used (whether extrusion, injection moulding or, for example, compression moulding) in that the inner structure - morphology, crystallinity, dispersion in a multi-phase system, orientation or internal stress pattern - predicts the properties (Michaeli, 1990). Our research work analyses injection moulding, extrusion, compression moulding, thermoforming and welding within this context. This paper is restricted to unfilled thermoplastics and to the first two processes mentioned. In all processes analyzed the structure formed is a function of thermal and mechanical (flow and deformation) influences. The absolute degree of structural properties, such as crystallinity or orientation, is different for the individual processes and the products formed but is always derivable from the specific process conditions. This gives us the opportunity to transfer findings from one production process to another.

The aim is therefore to abstract a process into individual parts for which the processing parameters are defined. From this the inner properties (structure) and final properties can be determined. This is shown in Figure 1 (Berghaus et al., 1988). Process data are figures independent of moulding and processing equipment. They are derived or calculated from the processing conditions and the specific geometrical boundary conditions of the equipment used, mould, die etc. Compared with compression moulding, for example, the material is more loaded (e.g. sheared, heated) during extrusion and therefore we have to expect molecular alteration. A comparison of extrusion and injection moulding gives us a similar set of significant process data, but the amount and effect are different. The pressure level and the shearing in injection moulding, for example, is significantly higher than in extrusion (Berghaus et al., 1988). The structural parameters (e.g. orientation or boundary layers) are less distinct in extrusion. During cooling in injection moulding

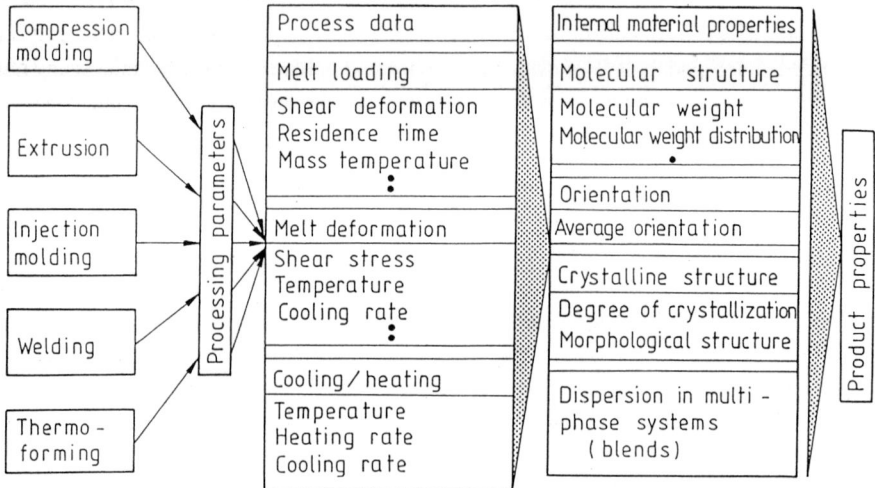

Fig. 1   From production process to part properties; from Berghaus et al. (1988).

comparatively the same cooling gradients exist, but due to the melt flow during the packing phase high deformation rates at fairly low melt temperatures influence the inner structure. The morphological structures and internal stress patterns can be correlated for both processes with the cooling, pressure and deformation conditions. Analysis of these fields provides the key to understanding the relations between processing and properties.

2. How to establish relations between processing and properties

Basically there are two different ways of establishing relations between the polymer and the processing on one hand and the resulting product properties on the other: the experimental method and the theoretical method by means of process and structure development modelling. Since this relationship is of an extremely complex nature most of the approaches found in the literature lie somewhere between the two methods but generally they are more experimental. Describing the flow and deformation with respect to the orientation of a polymer in a processing machine and its toolings and also its cooling and the parallel development of crystalline structures or internal stresses is no longer out of our reach when modelling polymer processing. Computers have proved to be essential for polymer engineers and scientists; in other words: simulation of polymer processing is already very advanced and, although it is definitely still a vision, the parallel prediction of part properties, including the mechanical properties, is becoming more and more realistic. We already see the day when it will be possible to the predict processing behaviour and properties of the product before processing, using adequate modelling (software) and computers (hardware). Examples of both these approaches will be given in the following.

3. Processing - property relations; the experimental approach

The state of a thermoplastics material while being processed and moulded to a plastics part is defined by (Berghaus et al., 1988):

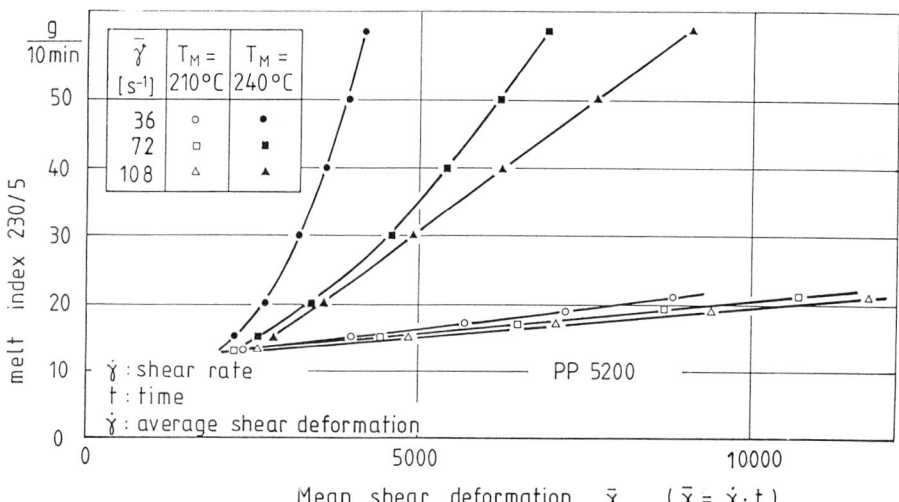

Fig. 2   Melt loading of PP; from Berghaus et al. (1988).

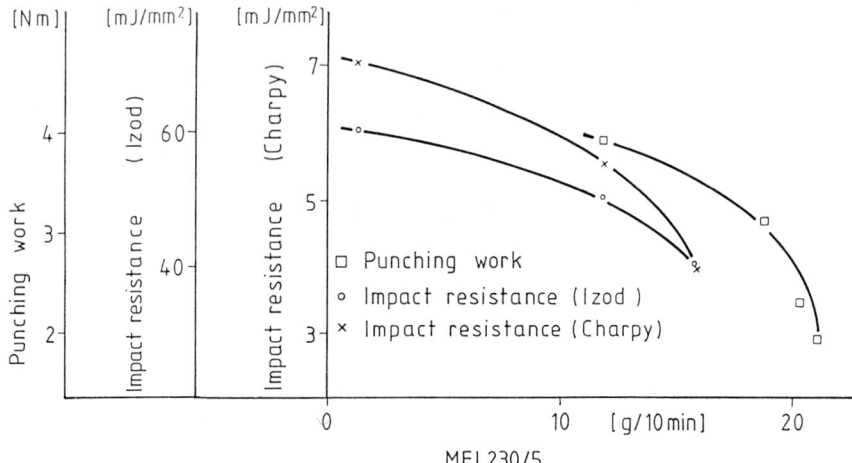

Fig. 3   Change in impact strength according to melt flow index; from Berghaus et al. (1988).

- melt loading (degradation),
- melt deformation (orientation) and
- cooling conditions

3.1 Melt loading and inner structure

To achieve product properties such as sufficient impact resistance, corrosion resistance and dimensional stability under mechanical loading a certain molecular weight is necessary. Molecular weight and flow properties (e.g. MFI measurements) can be correlated and give a good correlation with melt loading - i.e. degradation (Berghaus *et al.*, 1988). Figure 2 (Berghaus *et al.*, 1988) describes the effect of melt loading on MFI as a function of mass temperature and shear deformation. The significance of the thermal influence (mass temperature) can clearly be seen. This means that material processing might change the properties of our processed material. These experiments were carried out for extrusion (Hoffacker, 1982; Brinkmann, 1986); injection moulding shows the same tendencies.

3.2 Morphology and part properties

The changes in the inner structure of the material during processing ultimately determine the product properties. Tensile tests often do not reveal significant tendencies or correlations with the mo-

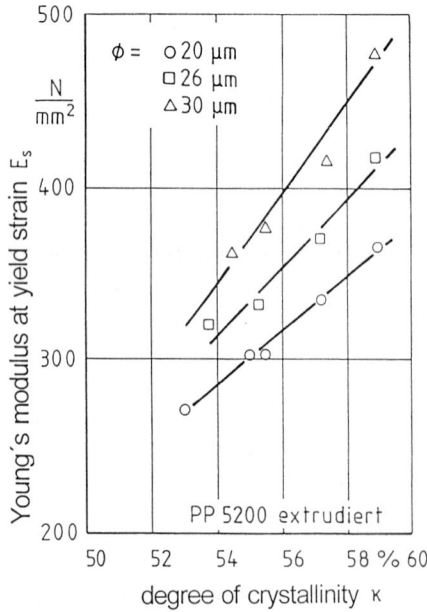

Fig. 4   Young's modulus at yield strain as a function of the internal properties; from Pleßmann (1986).

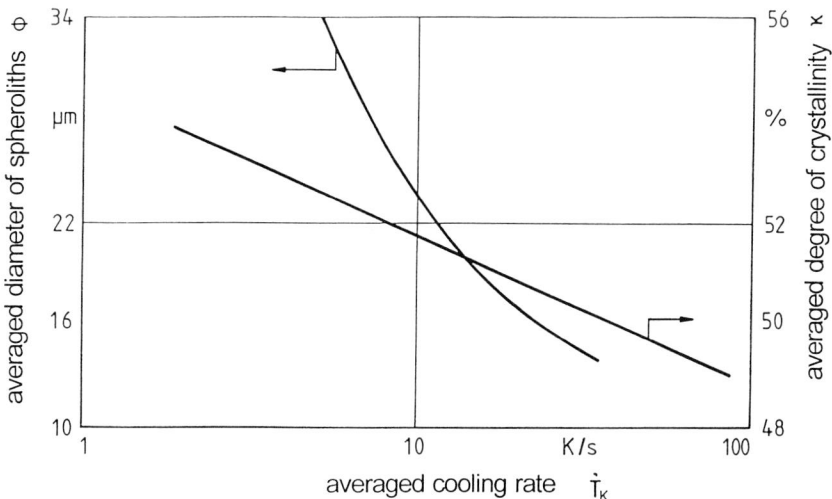

Fig. 5   Averaged diameter of the spherulites and degree of crystallinity as a function of the averaged coling rate; from Pleßmann (1986).

lecular degradation. However, if we analyse the material behaviour during impact loading (impact resistance, punching work) we can see from Figure 3 (Berghaus et al., 1988) how molecular, degradation (expressed here by means of the MFI) is of significant influence. Not only molecular degradation, for example in the screw or die of a processing machine, is of significant influence on the mechanical product properties. Cooling and quenching conditions are also very important.

Figure 4 (Pleßmann, 1986) shows how the properties (Young's modulus) of an extruded film from a semi-crystalline material (polypropylene) is influenced by the size of the spheroliths and the degree of crystallinity. It can be seen from the figure that knowledge concerning the initiation (nucleation), growth and influences on the number and size of the spheroliths is essential for the best processing conditions of that semi-crystalline material. The cooling rate is one of the key factors determining the mechanical properties (Figure 5; Pleßmann, 1986).

Figure 6 (Menges et al., 1988) shows the variation of the degree of crystallinity along the flow path for a number of different injected mouldings with simple geometries. For all these different geometries and processing parameters the mouldings show a marked variation in the degree of crystallinity along the flow path. According to Schönefeld and Wintergerst (Schönfeld and Wintergerst, 1970) a reduction in crystallinity of the order of magnitude shown here is expected to reduce the yield stress, Young's modulus and surface hardness by about 10 % and to increase the impact strength by the same amount. Microsampling measurements on moulded plates showed that the high average degree of crystallinity observed at the beginning of the flow path is due to the presence of a highly crystalline region near the surface of the moulding. This results from solidification under the influence of shear stresses, the effect of which becomes less with increasing flow distance. Altendorfer and Geymayer obtained similar results from investigations on injection moulded beakers (Altendorfer and Geymayer, 1983).

The thickness, degree of orientation, crystallinity and morphology of the region close to the surface, described by Backhaus (1985) as the freeze-off layer, determine to a large extent the mechanical properties of the moulding. If different mouldings made from the same material have identical structures at the positions where measurements are made, it can be assumed that the end properties will be similar. The mechanical properties measured in the tensile tests, namely Young's modulus E, the yield stress σ_S and the elongation at break ε_R, makes it possible to characterize the stress-extension behaviour of the materials under study and to make relative evaluations. The size of the tensile bars needed for tensile tests allows only the mechanical properties in the middle region of the moulding to be investigated. Flat compression tests, on the other hand, allow loading of the material even near the edges of the components being investigated, as the sample volume needed for the test is small. This makes it possible to determine the positional dependence of the mechanical properties at every part of the injection-moulded article (Menges et al., 1988). Samples were moulded from a PP homopolymer (®Vestolen P 5200, manufactured by Hüls, Marl). Plates with dimensions 114 mm x 114 mm and thicknesses of 1.5, 2.1 and 5.0 mm were made for the investigations. (Where the thicknesses given in the figures differ from these, the values given are actual thicknesses after shrinkage). Each test piece was injection-moulded with a film gate over the full width. The processing parameters which were varied during the investigation were the melt temperature T_M, the cavity wall temperature T_W, and the flow front velocity V_F. The holding pressure p and the corresponding time t_N, and also the cooling time t_C, were adjusted to suit the conditions in each case.

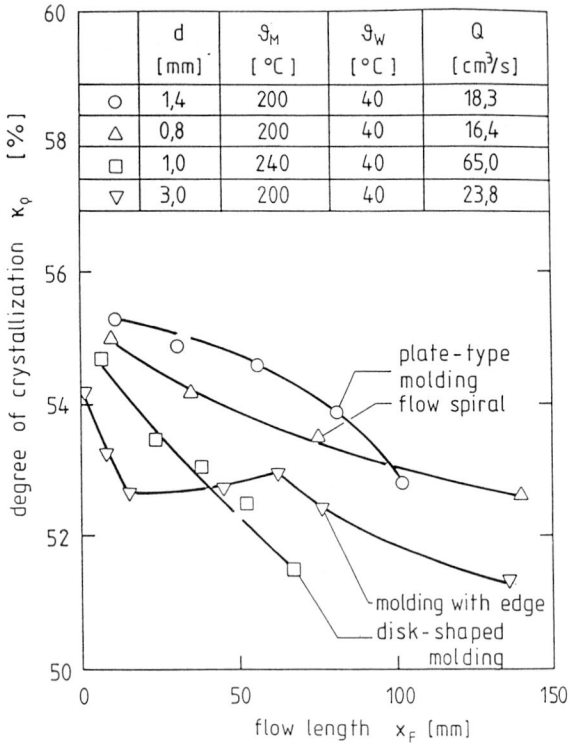

Fig. 6 Change of degree of crystallization along the flow path for different moulding geometries; from Menges et al. (1988).

The degree of orientation and the thickness of the fine-grained surface layer were determined at different positions along the flow path from 30 µm thick microtome cuts (Troost, 1985). The average degree of crystallinity over the cross-section was determined by the flotation method according to DIN 53479. The extent of the spherulite size over the cross-section and along the flow path was also determined, by point sampling (Wiegmann, 1987). These investigations showed that along the flow path the structure changes significantly only in the region near the surface. In contrast, the structure in the interior zone shows no noticeable variations. This indicates that the dependence of the mechanical properties on flow path length is determined essentially by the structure in the surface layer.

Flat tensile test speciments of small dimensions (unstretched length $L_0 = 25$ mm) were cut out of the mouldings in accordance with DIN 53455. The tensile tests were carried out at room tem-perature under the control of a plastometer, using a constant true extension rate of $\dot{\varepsilon}_w = 0,5$ min^{-1}.

The deformation behaviour under compression loading was determined by the flat compression test, Figure 7 (Menges et al., 1988). In this test two opposing rams with flat, well-lubricated working faces are pressed into an equally flat specimen. Under suitable geometrical conditions ($b/h > 6$) one can assume that the deformation of the specimen is confined to one plane and occurs in the directions I and II. The specimens

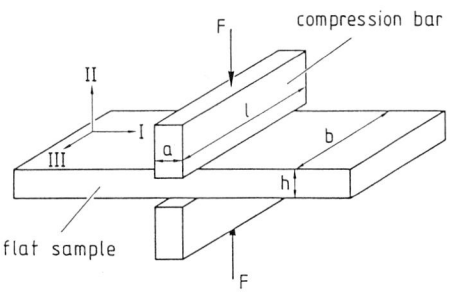

Fig. 7   Compression test; from Menges et al. (1988).

were cut out in such a way that the injection moulding direction coincided with either the direction I (sampling angle $\theta = 0°$) or the direction III ($\theta = 90°$). By choosing suitable specimen widths the ratio b/h was kept at 6, while the ratio h/a was kept at 1 (Menges et al., 1988). The tests were carried out at room temperature with a constant true compression rate $\dot{\varepsilon}_w = 0,5$ min^{-1}. The values measured in the test for the force F and the decrease in the thickness of the specimen were used to calculate the compressional stress $\sigma = F/(ab)$ for a ram width a and a specimen width b, and the percentage compression $E_t = [(h_0 - h)/h \cdot 100\%]$, where h_0 and h are the specimen thicknesses before and after applying the load respectively. In the following results the compressional strength is given as σ_{50}, which is the compressional stress corresponding to a compression $\varepsilon_t = 50\%$ (Menges et al., 1988).

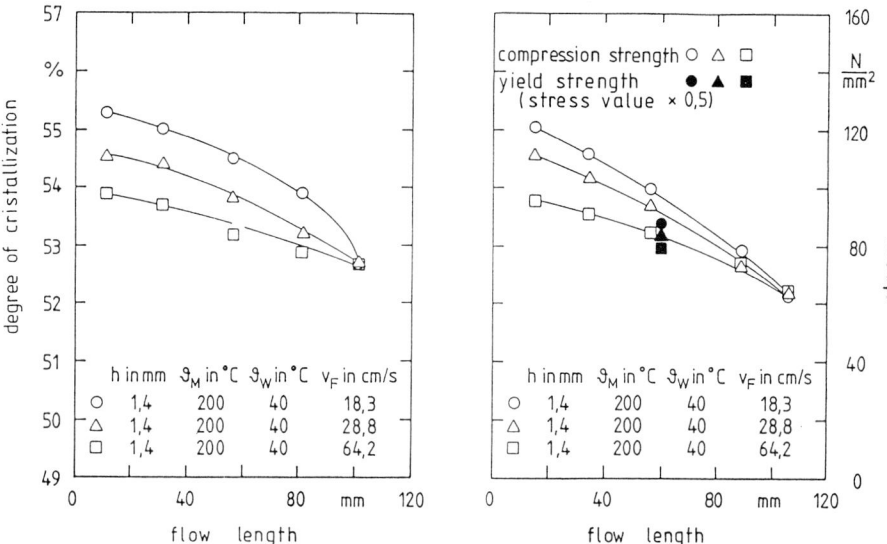

Fig. 8   Dependence of crystallinity and mechanical data on flow length; from Menges et al. (1988).

As already mentioned, for moulded components and semi-finished products made from semi-crystalline plastics the degree of **crystallinity** has a considerable influence on the mechanical properties. The magnitude of this influence was assessed by determining the yield stress σ_s and the compressional strength σ_{50}. Figure 8 shows that for a 1.4 mm thick plate, both the degree of crystallinity and σ_{50} change appreciably along the flow direction (Menges et al., 1988). As both characteristics have comparable dependences on the internal properties, the following figures show only the results from flat compression tests. The local flow velocity, which is almost identical to the flow front velocity V_F for the plates investigated here, has an appreciable effect on the properties mentioned above in the case of thin-walled mouldings. In general one finds that the dependence of the degree of crystallinity on V_F becomes less as the melt temperature is raised. The explanation of this is that at higher melt temperatures the effect of the flow front

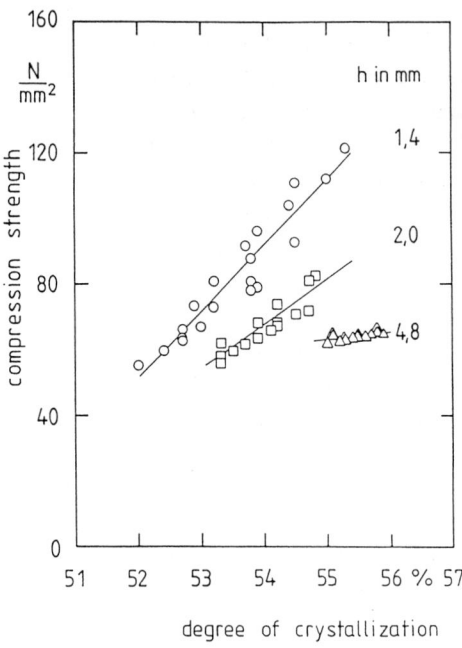

Fig. 9   Compression strength vs. degree of crystallization; from Menges et al. (1988).

velocity on the thickness of the surface layer is appreciably reduced. As already mentioned, the surface layer is highly crystalline owing to shear-induced crystallization, and consequently its thickness has a considerable effect on the degree of crystallinity when averaged over the cross-section of the moulding (Menges et al., 1988). With increasing thickness the dependence of both the structure and the mechanical properties on flow path length becomes less and less pronounced. The crystallinity, however, increases if the conditions near the wall are otherwise unchanged. The curves

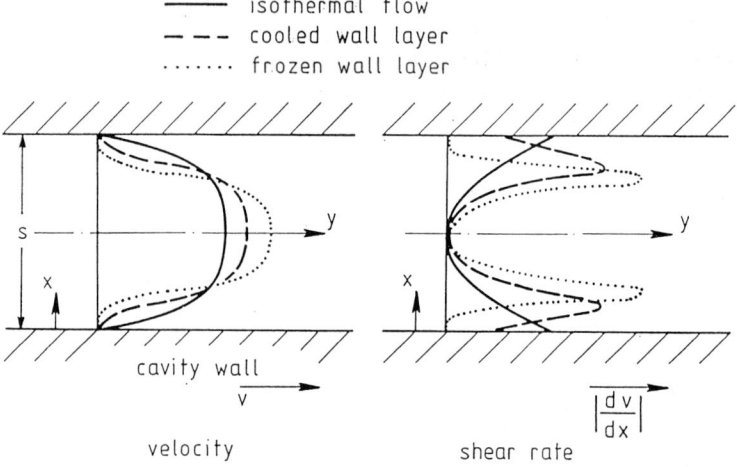

Fig. 10   Velocity and shear rate distribution across the flow channel at different points-of-time during the filling phase; from Wübken (1974).

shown in Figure 8 suggest that a relationship exists between compressional strength and degree of crystallinity. In Figure 9 these two quantities are plotted together for mouldings of three different thicknesses. For mouldings of equal thickness there is a very good correlation between compressional strength and degree of crystallinity. With increasing thickness it is found that for a given degree of crystallinity the compressional strength decreases (Menges et al., 1988). The compressional strengths measured for 4.8 mm thick plates correspond to those found at the end of the flow path for thin plates. This result shows that in addition to the degree of crystallinity there are other internal properties which have a considerable influence on the mechanical behaviour. Therefore the effects of orientation on the mechanical behaviour were also investigated (Menges et al., 1988).

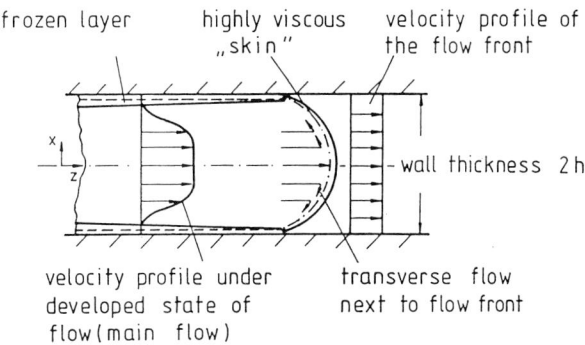

Fig. 11 Velocity profiles across the flow channel; from Wübken (1974).

Orientations develop during injection moulding mainly in the filling phase. Shearing and extensional deformation of the melt during flow in the runner system and the cavity lead to an orientation of the molecules in a preferred ("oriented") direction which is then frozen in during cooling. While the melt flows through a duct the highest shear rates occur next to the surrounding walls - if we presume wall adhesion (Figure 10; Wübken, 1974). The material cools down rather quickly at the cooled cavity wall and just about instantly a frozen layer is formed which again influences the flow pattern. This leads to a shifting of the maximum shear rate in the boundary layer between frozen material and the melt. Also during the filling phase the very flow front cools down to a highly viscous "skin" because of the cold air inside of the cavity. Due to the flowing melt this viscous "skin" is extensionally deformed like a balloon while being blown and is laid down at the cold cavity wall and freezes there instantaneously (Figure 11; Wübken, 1974).

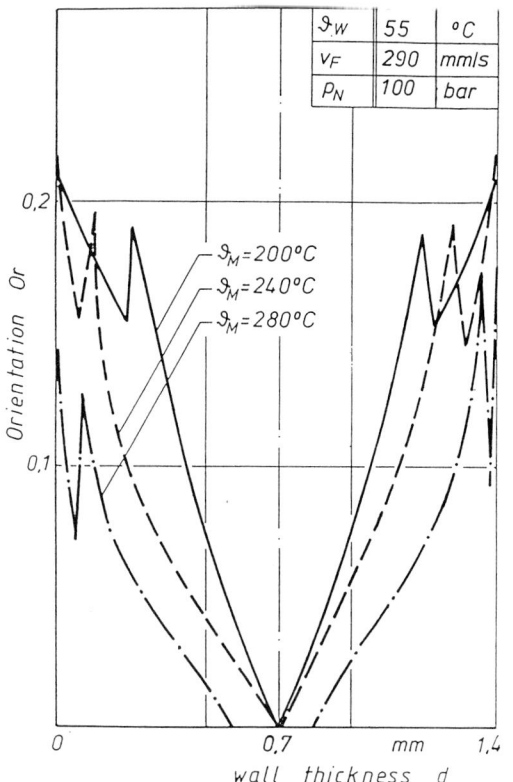

Fig. 12 Effect of mass temperature on the distribution of orientation; from Backhaus (1985).

In the final part the highest orientation will be found at such locations of maximum shear and extensional deformation. These orientations will relax with time according to an exponential function to reach their highest state of entropy since such orientations are of an entropy-elastic nature. With increasing temperatures these relaxation processes will be accelerated. Therefore we always find a certain relaxation procedure in a still warm moulded part (Figure 12; Backhaus, 1985).

When the dependence of the mechanical properties on position and direction was investigated, it was found that the behaviour of the material was highly anisotropic, to an extent which depended on the position in the moulded component. Figure 13 illustrates this for the compressional strength, measured at different positions of a plate with a thickness of 1.4 mm (Menges et al., 1988). On comparing the compressional strength values measured in the longitudinal and transverse directions at the same point in the moulding, quali-

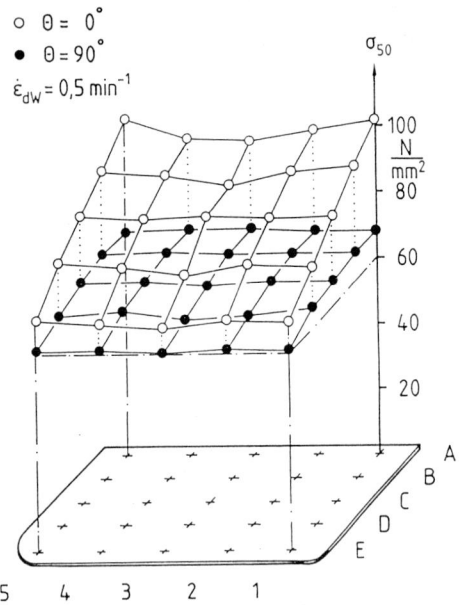

Fig. 13 Compression strength as a function of location in the moulding; from Menges et al. (1988).

tatively similar curves are found all along the flow path. In Figure 14 (Menges et al., 1988) the degree of orientation and the compressional strength are plotted against flow distance along the path marked 3 in Figure 13. In the figure the curves for two mouldings made under greatly differing conditions are compared. The compressional strength as a function of flow path length clearly shows the anisotropy which is expected as a consequence of the orientation, and also the effect of the change

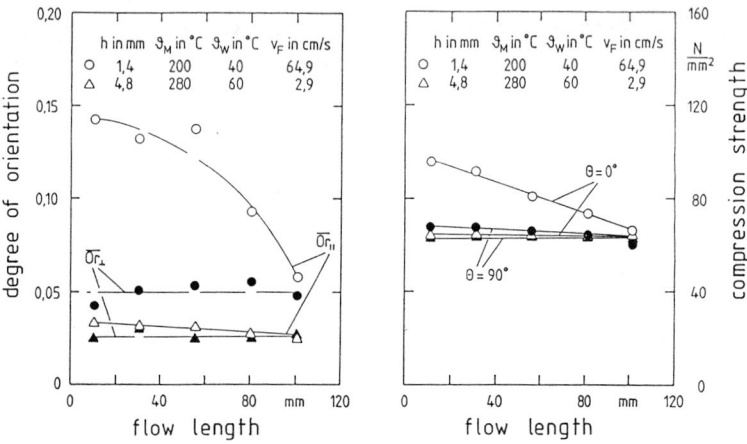

Fig. 14 Dependence of orientation and compression strength on flow length; from Menges et al. (1988).

in the degree of orientation along the path (Menges et al., 1988).

As can be seen from Figure 15 (Menges et al., 1988), in representing the compressional strength as a function of the average degree of orientation, generalized for mouldings of all kinds (or, to be more specific, without taking account of moulding thickness), the data show a large scatter. Nevertheless, unlike the situation found earlier for the dependence of compressional strength on degree of crystallinity, it is not necessary in this case to derive separate curves for different moulding thicknesses (Menges et al., 1988).

The observed scatter can essentially be attributed to two causes. The first of these is the uncertainties that arise in determining the degree of orientation by the birefringence method; according to Wiegmann (1985) the errors in this can be as much as 10 %. The second cause arises from the fact that the degree of orientation $\overline{O}r$ that is measured is an average value over the whole cross-section of the specimen. Although this gives an approximate measure of the amount of orientation present, equal values for the average degree of orientation can arise from either a thin, highly oriented surface layer, or a less highly oriented, but thicker, freeze-off layer. Although these different structures both influence the mechanical properties of the moulding in the same sense, the actual values of the effects are different (Menges et al., 1988). It may be possible from a more detailed characterization of the structure to reduce the amount of scatter in the dependence of mechanical properties on the degree of orientation. As the results show, use of the flat compression test makes it possible to demonstrate the positional dependence of the mechanical properties under compressional loading and to relate this to structural variations (Menges et al., 1988). By combining results from the compression test with the mechanical data generally used in designing structural components, it may be possible under some circumstances to arrive at more accurate formulas for predicting the behaviour of the material. The results analysed so far do not yet make it possible to fully attain this objective. Nevertheless, proper use of the flat compression test opens up new possibilities for predicting the mechanical properties of structural components.

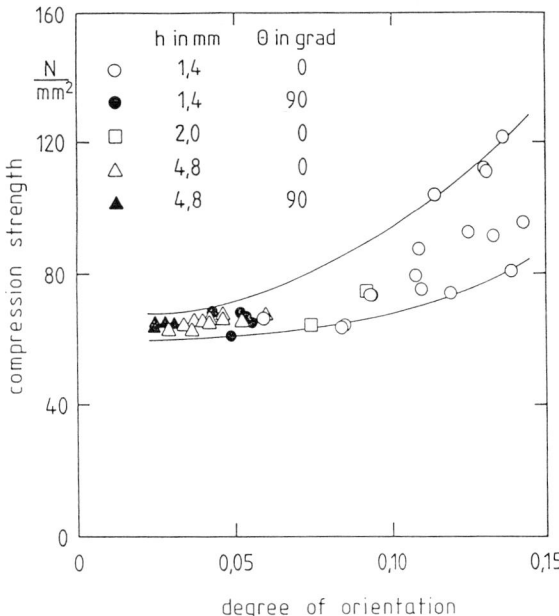

Fig. 15 *Compression strength vs. orientation; from Menges et al. (1988).*

4. Processing - property relations; the theoretical approach - a vision?

These experimental results show that the material properties are different at any position in an injection-moulded part. When we look at the overall part performance under loading conditions we should not neglect the internal stresses which develop during cooling and might negatively "preload" the part (Figure 16). These are difficult to detect; basically only physical modelling of their development during cooling and demoulding will help to show their magnitude and local distribution in a moulding and make it possible to judge their influence on the final part performance (Pötsch, 1991). In recent years significant advances have been made in order to model their development. A computer program for this purpose is already available for the injection moulding process (CADMOULD-3D, IKV, Aachen). At IKV we are trying to model the injection moulding process in order to predict the properties of the mouldings (Figure 17). In the first step we model the molecular orientation and the crystallization process.

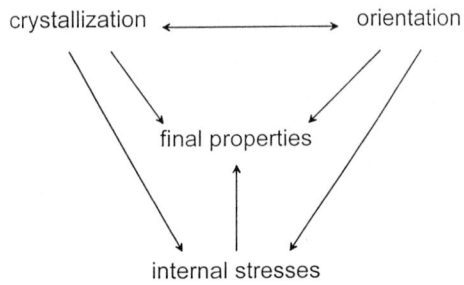

Fig. 16 *Relations between micro-structural parameters and final properties of the moulding.*

Orientation process

The orientation birefringence factor Δn is directly correlating to the normal stress difference Δs. If we succeed in predicting the orientation birefringence factor Δn out of the stresses acting up to the freezing of the material due to shearing and extensional deformation of the polymer at the flow front during filling, we have at least one significant key for the correlation of the orientations with the local mechanical properties. According to Tietz (1994), we have:

$$\Delta n = C \, \Delta \sigma \qquad (1)$$

where C is a constant and Δs is the main stress difference. For **orientation through shear** Δs is found from:

$$\Delta \sigma = \sqrt{N_1^2 + 4\sigma_{12}^2} \qquad (2)$$

where s_{12} is the shear stress, and N_1 the normal stress difference following from (Backhaus, 1985):

$$N_1 = \frac{2}{G} \sigma_{12}^2 \qquad (3)$$

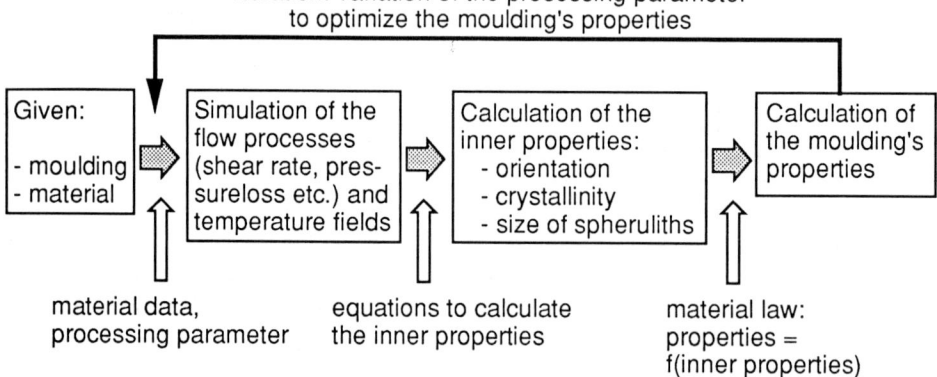

Fig. 17 *Simulation of the moulding's properties*

with G being the shear modulus. For *orientation through extension* the main stress difference in Equation (1) follows from:

$$\Delta\sigma = \eta_D \dot{\varepsilon}_F \qquad (4)$$

where h_D is the viscosity, and $\dot{\varepsilon}_F$ the extension rate at the flow front, which is obtained from:

$$\dot{\varepsilon}_F = \left(\frac{n}{n+1}\right)\frac{v_F}{\Delta X} \qquad (5)$$

in which n is the flow exponent, v_F the velocity of the flow front and $\Delta X = H/2$ (H being the height of the cavity).

Figure 18 shows a first qualitative calculation of the orientation pattern over the part thickness compared with a measured birefringence pattern. Figure 19 (Tietz, 1994) shows that there are still some deviations in the centre of the parts. The reason for this is that the modelling does not yet consider the melt flow during the packing phase of the injection moulding process; however, the qualitative function of the orientation - with the high orientation at the part surface according to the extensive molecular orientation at the flow front and the relative orientation maximum due to shear flow - is most encouraging. Figure 20 shows calculated orientations at different positions in a complex-shaped moulding (Tietz, 1994).

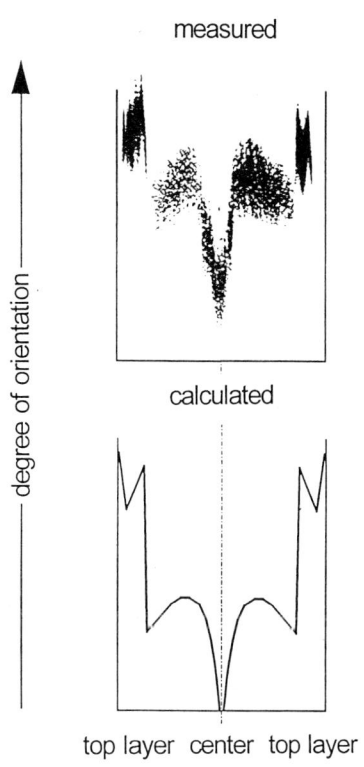

Fig. 18 *Comparison of measured and calculated orientation*

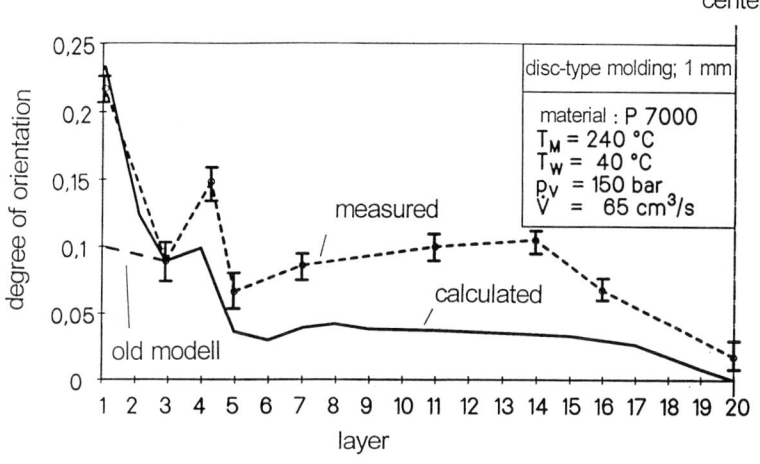

Fig. 19 *Comparison of measured and calculated orientation courses (close to the gate); from Tietz (1994).*

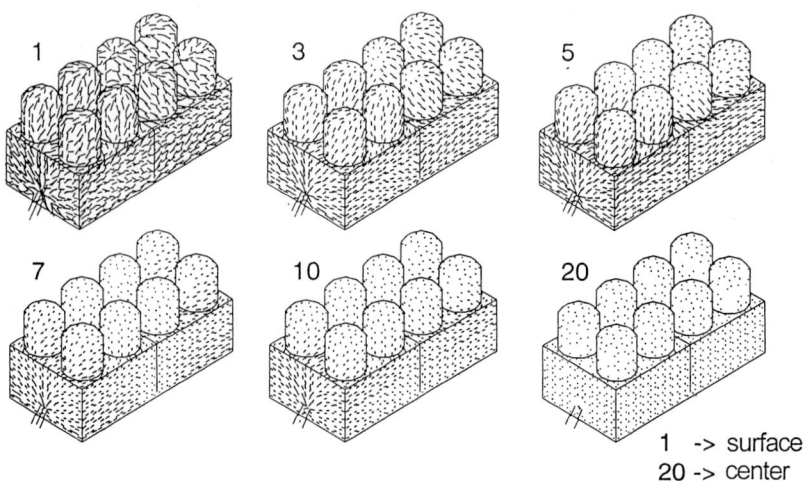

Fig. 20 Orientation distribution in different layers of a toy moulding; from Tietz (1994).

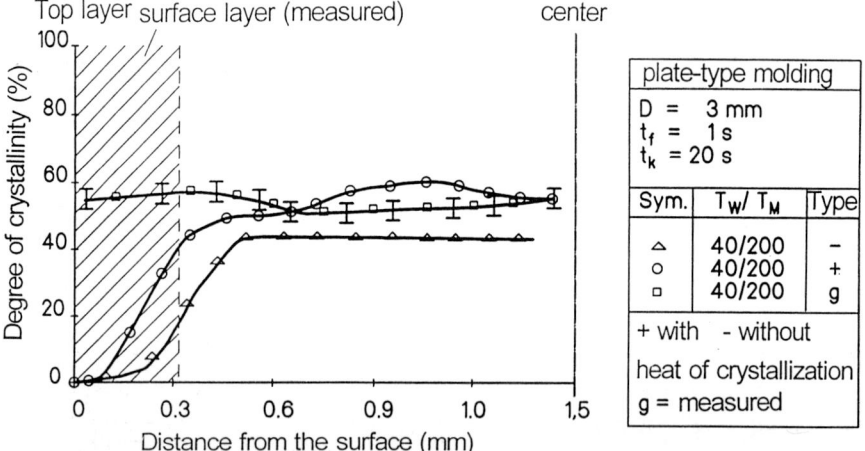

Fig. 21 Degree of crystallinity over the moulding thickness with/without heat of crystallization calculated and compared to measurements (close to the gate); from Tietz (1994).

Crystallinity

The Avrami equation for isothermal development of crystallinity is:

$$\chi(t) = \chi(\infty)\left(1 - e^{-Kt^n}\right) \tag{6}$$

where:
K = crystallization constant;
$\chi(\infty)$ = maximum crystallinity;
n = Avrami exponent ($n \approx 1...4$)

To describe the developing crystallinity of the moulding during cooling a non-isothermal formulation of the Avrami equation was integrated in our injection moulding computer program, resulting in (Tietz, 1994):

$$\chi(t) = \chi(\infty)\left(1 - e^{-\left(\int_0^t k(T(t))\mathrm{d}t\right)^n}\right) \quad (7)$$

with $k = K^{1/n}$. A comparison between the calculated and the measured degree of crystallinity distributions is given in Figure 21 (Tietz (1994). Since the shear-introduced crystallization close to the surface of the moulding is not yet integrated in our modelling, differences between the calculation and the measurement can be seen.

Assuming success in modelling and predicting the local crystallinity, local orientation and local internal stresses in a three-dimensional moulding, a relation to the local mechanical properties is still missing. There is still a large field for fundamental research. The Aachen University of Technology is working on this question together with the IKV team. Perhaps the first step in finding a relation is to use data as given in Figure 4 for prediction. However, this will only give information for the local modulus; predictions of failure behaviour are certainly even more complex.

5. Final remarks

The calculation of processing and mechanical properties by computer modelling is at the moment still a vision. However, when we analyse the processing of short fibre-reinforced polymers it is already possible to calculate fairly exactly the local fibre orientation and the fibre distribution even in very complex moulding (Figure 22; Mohr-Matuschek, 1991).

Since the modulus of the fibres (e.g. glass fibres) is much higher than that of the polymer we can apply the rules we know from the mechanical analysis of composite materials ("laminate theory") in order to predict the overall mechanical performance of the part (Mohr-Matuschek, 1991). This brings us very close to our vision!

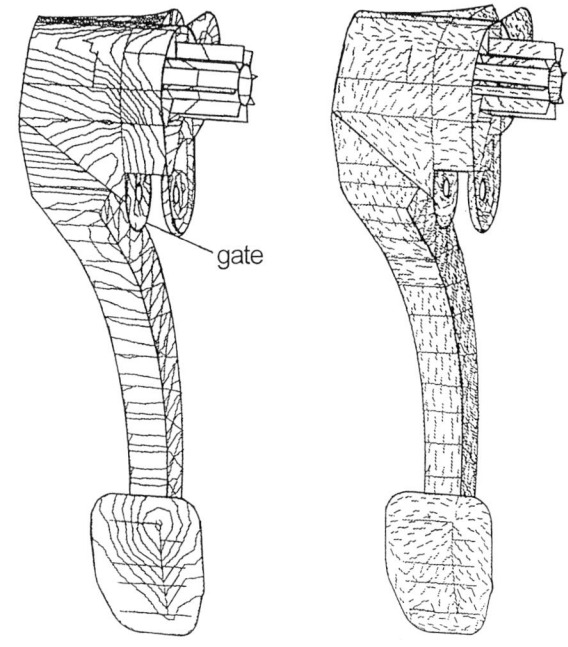

Fig. 22 Filling simulation for the clutch-pedal geometry; from Mohr-Matuschek (1991).

References

Altendorfer, F., Geymayer, G. (1983), Werkstoffentwicklung für dünnwandige spritzgegossene Verpackungsteile. *Plastverarbeiter*, 34 - 6, 511 - 514

Backhaus, J. (1985), *Gezielte Qualitätsvorhersage bei thermoplastischen Spritzgießteilen*. Dissertation, RWTH Aachen, Aachen, Germany.

Berghaus, U., El Barbari, N., Offergeld, H., Pötsch, G. and Ries, H. (1988), Material-Machine- End Product - Internal Structure as the key to component properties. *Preprints 14th Kunststofftechnische Kolloquium*, IKV Aachen, Aachen, Germany.

Brinkmann, T. (1986), *Karakterisierung des Einflusses der Schmelzebelastung auf den Zustand extrudierter Tafeln*. Studienarbeit IKV Aachen, Aachen, Germany.

Hoffacker, D. (1982), *Qualitätsbeeinflussung von Polystyrol- und Polypropylenschmelze*. Diplomarbeit IKV Aachen, Aachen, Germany.

Menges, G., Troost, A., Koske, J., Ries, H. and Stabrey, H. (1988), Fließwegabhängigkeit von Struktur und mechanischen Eigenschaften, *Kunststoffe*, 78 - 9, 806 - 809

Michaeli, W. (1990), Polymer Processing - Process Conditions, Structure Development and Final Component Properties. *5th Polymer Processing Society Meeting*, Kyoto, Japan.

Mohr-Matuschek, U. (1991), *Auslegung von Kunststoff- und Elastomerformteilen mittels Finite-Elemente-Simulationen*. Dissertation, RWTH Aachen, Aachen, Germany.

Pleßmann, K.W. (1988), Korrelation von Fertigung und Bauteileigenschaften bei Kunststoffen. *Ergebnisbericht des SFB*, 106.

Pötsch, G. (1991), *Prozeßsimulation zur Abschätzung von Schwindung und Verzug thermoplastischer Spritzgußteile*. Dissertation, RWTH Aachen, Aachen, Germany.

Schönfeld, G., Wintergerst, S. (1970), Beeinflussung von Struktur und Festigkeit von Polypropylen durch Wärmebehandlung, *Kunststoffe*, 60 - 3, 177 - 184

Tietz, W. (1994), *Modelle zur rechnerischen Vorhersage der Mikrostrukturausbildung in thermoplastischen Spritzgußteilen*. Dissertation, RWTH Aachen, Aachen, Germany.

Troost, A. (1985), Korrelation von Fertigung und Bauteileigenschaften bei Kunststoffen. *Ergebnisbericht des SFB*, 106, 133 - 153

Wiegmann, T. (1985), *Morphologische Untersuchungen an spritzgegossenen Polypropylen-Winkelformteilen*. Studienarbeit, IKV Aachen, Aachen, Germany.

Wiegmann, T. (1987), *Ausbildung der Gefügestruktur in Abhängigkeit von thermischen Randbedingungen*. Studienarbeit, IKV Aachen, Aachen, Germany.

Wübken, G. (1974), *Einfluß der Verarbeitungsbedingungen auf die innere Struktur thermoplastischer Spritzgußteile unter besonderer Berücksichtigung der Abkühlverhältnisse*. Dissertation, RWTH Aachen, Aachen, Germany.

*P. E. J. Flewitt**

Structural Integrity Assessment of High Integrity Structures and Components: User Experience

Reference: Flewitt, P.E.J. (1995), Structural Integrity Assessment of High Integrity Structures and Components: User Experience. In: *Mechanical Behaviour of Materials* (ed. A. Bakker), Delft University Press, Delft, The Netherlands, pp. 143-164.

Abstract: The continued safe operation of high integrity structures and components is assured using structural integrity arguments which are usually based upon deterministic methodology. In this review the methodologies used to develop these arguments are considered with emphasis given to the conservatisms in the input parameters, particularly the mechanical and physical properties of the materials. An example of a multi-legged deterministic fracture mechanics based argument is addressed by describing that formulated to support the continued operation of the high integrity reactor steel pressure vessels at four carbon dioxide gas cooled Magnox power stations operated by Nuclear Electric plc (UK). Finally there is a brief consideration of the potential benefits associated with probabilistic arguments compared with a deterministic approach.

1. Introduction

Major construction, process, manufacturing and transport industries are usually concerned to be able to assure the safe and economic production by use of appropriate structures and components for operation of their plant. An essential input to demonstrating this necessary assurance is a structural integrity assessment. The design of these structures and components is an interactive process which aims to achieve a realistic balance between state-of-the-art structural capability and the service requirements. This balance is achieved by use of codes and standards, methods of analysis, material property databases and validation tests. As a consequence the design engineer must create a structure or a component using analytical tools and supporting data most appropriate for the application. However, in addition, the operator of the structure or component has to ensure that throughout the service life it is secure against the design intent. This latter aspect offers challenges with respect to structural integrity diagnosis and assessment. In all such cases it is essential to have a knowledge of the mechanical and physical properties of the materials used to construct these structures and components.

In general the probability of failure with time in service for structures and components follows a curve of the form given in Figure 1, Neubauer and Bietenbeck (1989). This figure shows that during the early part of the operating life, there is a greater probability of failure which decreases and then rises slowly as the design life is approached, but then increases rapidly. Problems encountered during commissioning and the early life usually emerge as a result of poor design, inadequacies in component fabrication, construction and assembly, whereas in the latter stages of life they are associated with degradation of the properties of materials and the onset of time dependent failure mechanisms. However, in the intermediate stages, problems can develop, at varying levels, due to the unforeseen response of the material to the service environment, specific requirements of the plant operating regime, inadequacies in the manufacturing route and poor maintenance, Endo (1994),

---

* *Nuclear Electric plc, Engineering Division, Berkeley Technology Centre, Berkeley, Gloucestershire GL13 9PB, United Kingdom*

Swift (1994). An extreme example is the well documented catastrophic failure in 1969 of a low pressure steam turbine disc, 1.5m diameter, in a turbine generator at a power station in the UK, Figure 2(a), Grey (1972), Kalderon (1972). This was a result of the initiation and growth of stress corrosion cracks within the bore of the discs, Figure 2(b), arising from the operating steam environment combined with a material, in this case a low alloy 3CrMo ferritic steel, with a low fracture toughness, ≈40 MPa√m. The consequential effects of this are shown in Figure 3 where, although there were no nuclear safety implications, the costs to return the turbine generator to electrical power generation were substantial. Here it was the more onerous environmental service conditions during periods off load not addressed in the design, combined with poor materials properties for the particular rotor discs that produced the premature failure.

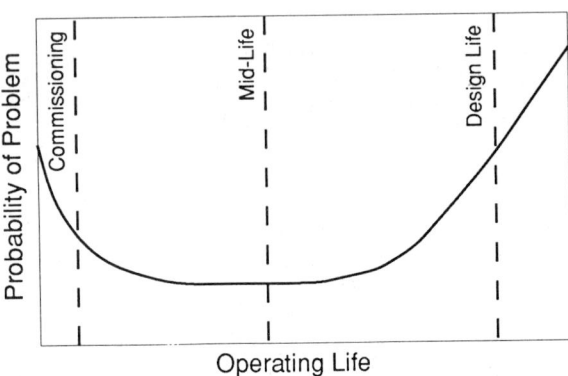

Fig. 1   The change in the probability of the incidence of structural integrity problems in relation to the operation life of plant

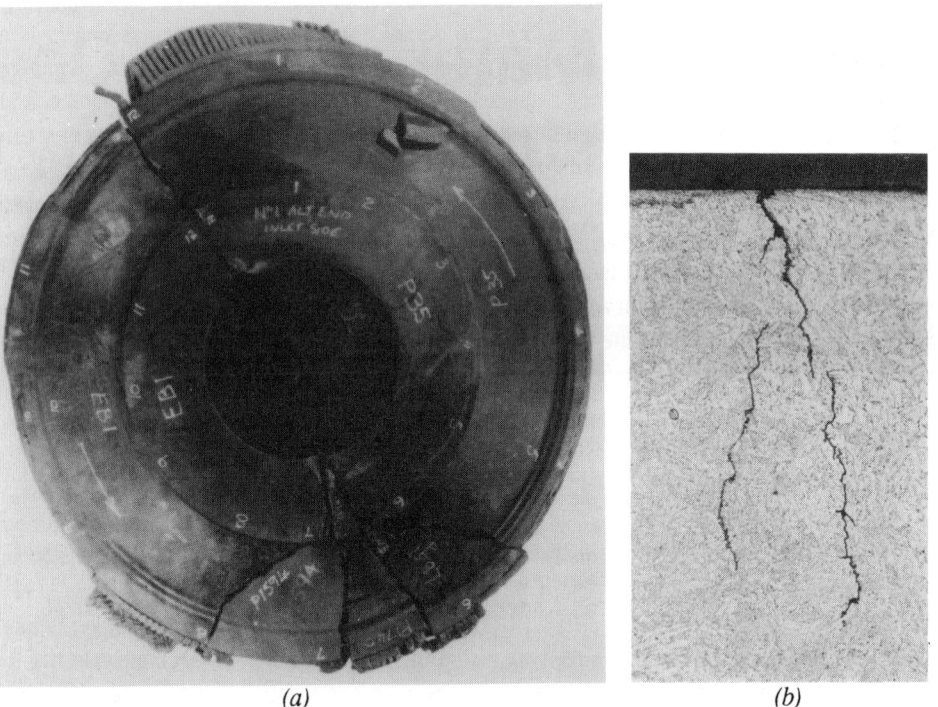

Fig. 2   Failure of a low pressure stage steam turbine disc manufactured from a low alloay ferritic steel. (a) Reassembled pieces of the 1.5 m diameter disc which failed while rotating at 50 Hz. (b) Failure was caused by the growth of a small SCC crack of the type shown here in a steel of low fracture toughness.

Fig. 3 A general view of the consequential effects of the low pressure steam turbine disc failure shown in Figure 2.

As a consequence it is essential that materials selected for high integrity service structures and components have a high measure of reliability in their properties. If consistent performance is not guaranteed other factors such as materials selection costs, overall economic service life etc., cannot be evaluated. Consideration of reliability has to focus on the potential failure modes and how well these can be predicted. Certainly it is usual to be able to diagnose the reason for failures after they have occurred. However the prediction of service life is less secure and as a consequence the design has to accommodate uncertainties by adding conservatisms and supporting these for high integrity plant with use of surveillance specimens, monitoring the operating conditions the component experiences and non-destructive inspection. These are used either individually or more usually in combination against a declared strategy. As pointed out by Prager (1991), the challenge is to develop predictive models for life combined with mechanical property data to support these. The latter offers considerable challenges since components are usually of larger cross-section than the test specimens and the environment and stress state more complex. In addition, the period of service required for the component usually exceeds that achievable in a laboratory test, so that acceleration in parameters including stress, temperature and environment is required combined with extrapolation to the service conditions. Thus there is a need to ensure that the mechanisms associated with the accelerated tests remain those experienced in service. Here there are benefits associated with being able to develop realistic mechanistic maps, for example, deformation and fracture maps, Figure 4, extending to stresses, temperatures and environments which are appropriate to service, to ensure the operating mechanisms are accommodated (Cocks and Ashby 1982). Here we show such a map for α-iron indicating the temperature and stress ranges over which deformation mechanisms apply. These maps could, and have in some cases been developed for given materials but there is a need to accommodate the specific composition, heat treatment, microstructure etc. which is difficult, but if available this greatly assists such extrapolations.

Obviously a failure of a structure or component can have profound commercial and financial implications. However, there are areas where failures are completely unacceptable due to the impact on the environment and society, and this is particularly relevant to the chemical, transportation and nuclear industries. As a

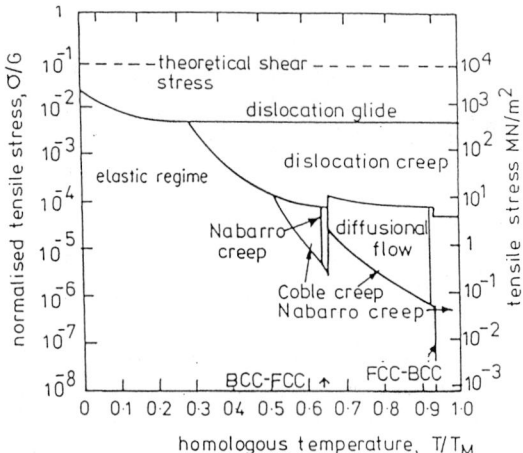

Fig. 4 An example of a mechanism map for α-iron (grain size 32 μm) describing the stress and temperature regimes which can be used to provide guidance for extrapolation to service conditions for structures and components.

consequence, in the UK the reliable, safe and economic continued operation of nuclear plant, including electrical power generating plant, is assured by the implementation of planned maintenance schedules supported by an appropriate range of safety arguments. To specify the maintenance schedules and formulate the safety arguments for electrical power generating plant requires a comprehensive understanding of the capability of the plant to meet the required duty based upon a structural integrity assessment. The commercial benefits to be obtained from structural integrity assessments associated with plant breakdown and continued operation beyond the original design life have been described previously, Doig and Gasper (1992) and Flewitt and Williams (1994).

An unplanned plant breakdown, or a fault identified during a scheduled outage, results in loss of availability or capability so that it is necessary to utilise structural integrity assessments. To meet the UK regulation requirements these arguments are usually formulated into a Safety Case to:

(i)   allow continued operation with known faults,

(ii)   establish appropriate plant monitoring,

(iii)   define a repair or replacement strategy,

(iv)   define acceptance criteria for continued operation.

As a consequence, it is possible to minimise loss of income and additional expenditure by implementing a Safety Case, effecting a repair or replacement, or providing an operational solution, Figure 5(a). Here is shown schematically the cost and time associated with an unscheduled outage together with the effect on the total operating cost, as designed, and the revised income. This indicates the reasons to minimise period and cost of the outage which would arise from the need to undertake an assessment and prepare a Safety Case, effect a repair, or replace a component. The financial break-even point for the plant is shifted in time, thereby decreasing overall profitability to

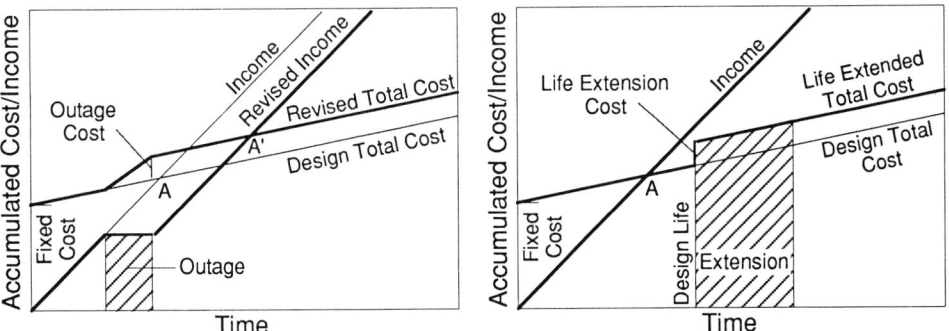

Fig. 5 The economics associated with operating electrical power plant including the initial fixed construction cost for (a) an outage (A = Design life, financial breakeven point; A' = revised financial breakeven point); (b) extension beyond the design life.

the point where, in the extreme, continued operation may not be economic. These financial losses have to be recovered subsequently by improved operating efficiency, possibly combined with an extension to the operating life of the plant. This is supported by a structural integrity assessment to establish life limiting parameters, improving the knowledge of the operating history, identifying conservatisms in the original design and testing the design against acceptance criteria. The economic benefits associated with continued operation of plant beyond the design life are illustrated in Figure 5(b) where there is a need to undertake work to accommodate the increase in problems as shown in Figure 5(a). However, for this strategy to be effective the life extension cost has to be minimised.

In this review paper the methodologies used to develop the structural integrity arguments for assessing high integrity structures and components are considered by reference to that adopted in the UK for nuclear electricity power generating plant. Section 2 describes the framework for formulating these structural integrity arguments. In Section 3 the arguments required to support the continued operation of reactor steel pressure vessels used at four gas cooled Magnox power stations operated by Nuclear Electric plc are considered. Finally an overall approach to the deterministic assessment of high integrity plant is considered with respect to a probabilistic argument in Section 4.

2. Basis of Structural Integrity Arguments

Structural integrity assurance is the demonstration that a structure or component meets its required duty with appropriate consideration of safety and economics. As a consequence, it is a multi-discipline activity which involves inspection, diagnosis, assessment and formulation of safety arguments, design criteria and repair/ replacement strategies and their implementation. It draws mainly from the technical disciplines of:

(i)   inspection, plant monitoring and remote handling,

(ii) material science including materials properties, fracture mechanisms, welding technology and tribology,

(iii) structural analysis including stress analysis, dynamics, continuum mechanics and fracture mechanics, and

(iv) general engineering safety and economic assessment.

Certainly it applies to all stages of the life of a structure or component, from conception to decommissioning, and as such it is not simply, as is often considered, a failure and fault diagnosis procedure. Moreover it has to be supported by appropriate research and development work. The structural integrity inputs to a particular problem are multiple and varied and, as shown in Figure 6, they involve the key elements of:

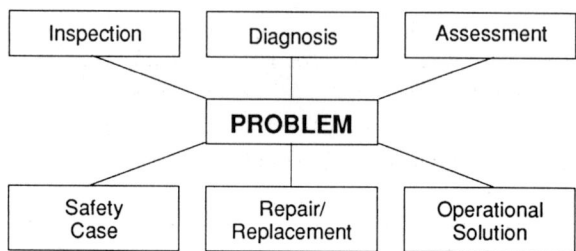

Fig. 6 The key elements for a structural integrity assessment which leads to a safety argument, a repair or replacement or an operational solution to a problem

(i)   Inspection   (iv) Safety Argument Formulation

(ii) Diagnosis   (v) Repair or Replacement

(iii) Assessment   (vi) Operational Solution

The first three provide the necessary input to the latter three and, for a given problem, each will have an input, although the weighting assigned will vary with the nature of the particular problem. In the case of the inspection, diagnosis and assessment of a problem, the inputs span the plant investigation work at the site, to longer term research and development activities to provide the necessary tools to solve the problem. In the case of inspection, however, it is to provide the hardware for techniques of inspection together with the interpretative capability to support the detection limits and resolution for defect sizing. In addition, for undertaking work within a nuclear reactor circuit, for example, it is necessary to develop both the appropriate delivery systems and optimise the techniques for remote operation. For problem diagnosis, work is required to offer the correct understanding of the mechanisms leading to any deterioration of the plant integrity or potential failure to allow the correct structural integrity assessment, safety argument, repair or replacement strategy or operational solution to be adopted.

As considered by Wannenberg, Klintworth and Rauth (1992) a deterministic structural integrity assessment has to be conservative and of sufficient accuracy as shown schematically in Figure 7. Here a simplified analysis adopts conservative assumptions whenever an uncertainty exists, to give a conservative assessment of safety. If the simple structural integrity assessment gives a level of safety that would require, for example, a costly repair or replacement the option remains to undertake a more detailed analysis which could lead to less conservative, but sufficient, results. Therefore such an approach is an option in formulating a deterministic

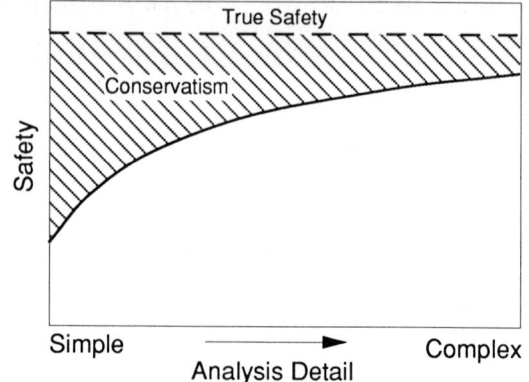

Fig. 7 Schematic diagram showing the amount of analysis detail required with respect to the level of safety associated with a structural integrity assessment.

argument, although the true safety level is unaltered. Unfortunately, it is not usually possible to establish if a more detailed analysis provides an appreciably higher estimate of the safety of the structure or component since the true safety level is unknown. A simplified analysis should always provide a more conservative assessment than a detailed analysis. However, several of the parameters used in a deterministic fracture mechanics assessment are probabilistically distributed variables. Materials properties exhibit scatter, defects and their sizes are statistically variable and loadings may be random. Thus a deterministic assessment assumes the lowest value options of the variables to give a conservative and, indeed, often over-conservative assessment. As a consequence, a probabilistic approach offers the potential to incorporate the appropriate level of detail into the analyses to lead to a more realistic assessment of the level of safety. The developments that lead to probabilistic approaches have been described elsewhere, and will be considered further in Section 4, but we will concentrate on deterministic approaches.

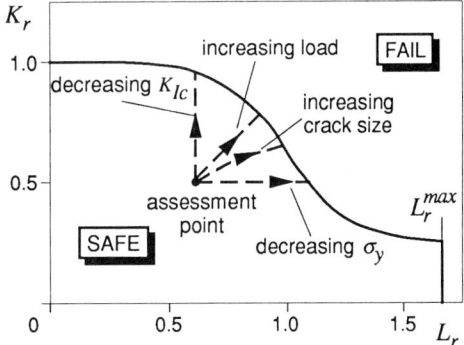

Fig. 8 A typical R6 failure assessment diagram showing an assessment point within the safe region, together with the trajectories of this point when fracture toughness K_{Ic}, yield stress σ_y, load and crack size are varied.

Although each deterministic assessment could be carried out in an ad hoc manner, the use of standardised procedures leads to greater efficiency and improves the acceptability to the Regulator. As a consequence significant effort was devoted to research and development within the CEGB, and more recently Nuclear Electric, to this approach. The R6 procedure Milne et al (1986) is well recognised and established for assessing structures which potentially could fail by any combination of fracture and plastic collapse. This procedure introduces the concept of a failure assessment diagram to link linear elastic fracture mechanics, K, and more recently fully plastic, J, solutions with fully plastic fracture concepts, Figure 8. To use this procedure the severity of loading on a structure containing a defect or an assumed defect is calculated in terms of two parameters (i) L_r which is proportional to the plastic yield load and (ii) K_r which is proportional to the load for LEFM failure. The two parameters are plotted as a co-ordinate pair on the diagram, Figure 8, and if the assessment point lies outside or on the assessment line possible failure is identified. Also included here is the effect of certain input materials properties and parameters on the trajectory of the assessment point.

This procedure is used routinely as a basis for deterministic fracture mechanics safety arguments supported by British Standards and ASME documentation. The success of this approach has spawned the development of a family of R procedures, to address the response of structures to unsteady loads, impact events, seismic loadings and high temperature loadings. In addition, the concern for safety is addressed by a consideration of the consequences of structural failure to the general public, the environment and the operator where the basic principles are defined taking into account the 'Safety Assessment Principles for Nuclear Plant' issued by the Regulator, H M Nuclear Installations Inspectorate (1992). As a consequence, the Safety Standards adopted in the UK by Nuclear Electric are monitored by Company Assessors and the Regulator who have to be satisfied with compliance to an agreed level of safety. This requires that structural integrity Safety Cases are formulated on the basis of the technical inputs of inspection, diagnosis and assessment. In general, a Safety Case considers various inputs each of which forms a leg of a multi-legged argument where the overall argument is judged on the basis of the strengths and weaknesses of each leg.

An alternative to a safety argument is to repair or replace a component where this has a clear economic benefit over the operating life of the plant. For example, in the case of repairs which involve welding, specific procedures are developed and underwritten by appropriate research and development work, although on many occasions it is sufficient to use the results of a welding trial or procedure test. Moreover, in the case of a repair or replacement, the integrity has to be underwritten for future operation. The arguments described for safety cases and repair or replacement solutions apply equally to operational solutions which are embodied in rules and instructions against which the plant is then operated.

3. A Deterministic Structural Integrity Assessment Case History

In this section consideration is given to the structural integrity assessment of Magnox reactor steel pressure vessels which are used in electrical power stations operated in the UK by Nuclear Electric. These provide a demonstration of the underlying philosophy and approach adopted for a rigorous, deterministic fracture mechanics based argument.

3.1. Background

The four Magnox power stations with reactor steel pressure vessels which are operated by Nuclear Electric plc were designed in the early 1960s against a requirement that failure in service should be incredible. Therefore, the vessels were designed and manufactured to the highest standards of the period, Holliday and Noone (1961), Poynor (1969), and procedures were put in place to ensure that they were not operated outside their design limits. Since construction all the vessels have operated satisfactorily and changes to the properties of the steels used in their fabrication as a result of exposure to neutron irradiation and service temperatures have been monitored via in-reactor surveillance schemes. The surveillance scheme results have been used as a basis for periodically reviewing the case for continued safe operation of the vessels taking into

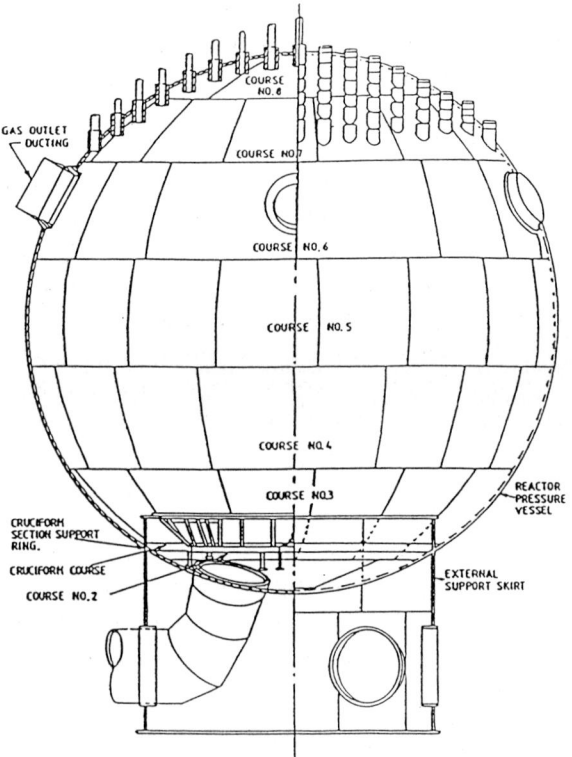

Fig. 9 Schematic diagram of a Magnox reactor steel pressure vessel showing distribution of plates and welds together with the inlet and outlet gas ducts.

Table 1 Typical chemical compositions of reactor pressure vessel steel plate and forging, submerged arc weld metal, and manual metal arc weld metal (wt-%)

C	Si	Mn	P	S	Cr	Mo	Ni	Al	As	Cu	Sn	Fe
Plate												
0.15	0.17	1.15	0.02	0.035	0.03	0.01	0.06	0.057	0.018	0.10	0.012	Bal.
Nozzle Forging												
0.18	0.36	1.30	0.024	0.024	0.07	0.01	0.08	0.049	0.029	0.10	0.015	Bal.
Submerged Arc												
0.07	0.72	1.80	0.035	0.051	...	0.025	...	...	0.04	0.25	0.02	Bal.
Manual Metal Arc												
0.08	0.45	0.90	0.025	0.027	0.06	0.02	0.06	<0.01	0.032	0.06	<0.01	Bal.

account any new developments in fracture mechanics which have occurred since the vessels first entered service and the requirement to operate beyond their original design lives. The Safety Cases developed for these major components provide an example of the formulation of a multi-legged structural integrity safety argument.

The Magnox reactor pressure vessels are essentially 20m diameter spheres fabricated from 75 to 100mm thick carbon steel plates and forgings jointed together by either manual or machined made welds. The composition of these steels are given in Table 1. A schematic diagram showing the typical distribution and form of welds and plates is presented in Figure 9. This spherical shell contains the carbon dioxide coolant gas and operates, typically, at a pressure of 1.8 MPa and a temperature of $\leq 340°C$. As with any structure of this type it is the welds that are judged to be the areas of the vessel most likely to contain any significant defects. At the time of their construction in the early 1960s there was no specific nuclear code available and the vessels were designed, constructed, inspected and tested to the conventional code BS1500 Class 1. Thus, the original safety assessments for the vessels were based on the fact that they had been built to an established code of practice taking particular care in the quality control, fabrication and inspection techniques. Although BS1500 did not directly address the problem of brittle fracture, the designers recognised the risk from this mode of failure and took steps to eliminate concern by establishing Operating Rules which prevented significant pressurisation of the vessel until the minimum operating temperature exceeded some limiting value. This limiting value, derived on the basis of a crack arrest philosophy, was obtained by combining the crack arrest temperature measured on plate material, with an upward shift in this temperature to accommodate in-service neutron irradiation and thermal ageing effects. Wide plate tests on full thickness specimens taken from similar steel plates indicated crack arrest temperatures which were typically 30°C. A nominal 40°C shift was then added to allow for in-service degradation. Specimens of the pressure vessel steels were installed in canisters within the reactors to monitor the extent of this temperature shift and hence provide a basis for any necessary future modification to the Operating Rules.

During the subsequent period of operation, tensile and Charpy impact energy specimens have been withdrawn from the surveillance schemes for testing on a regular basis. By the late 1970s these tests indicated that the allowance for irradiation/ thermal ageing effects used in setting the original Operating Rules had been exceeded and that some revision of these was necessary. The largest change in material properties occurred in specimens of machine-made submerged arc weld metal for which the upward temperature shift in the 40J Charpy energy was greater than 40°C. The change in materials properties for plate, forging and manual metal arc weld specimens were considerably less. This difference in response to neutron irradiation was attributed to the higher copper content of the submerged arc weld metal resulting from the coating on the original weld consumable, Table 1.

As a consequence of the results from the monitoring schemes, the integrity of the pressure vessels was reviewed in the early 1980s and the safety arguments revised using a fracture mechanics methodology. Since that time the safety arguments have evolved to a high level of sophistication and further modifications to the vessel operating procedures have been made. The assessments carried out are now supported by the results from an ultrasonic inspection of a sample of vessel welds at two power stations, Bowring et al (1993), Curry and Burrows (1993).

3.2. Basis of the Assessment

The basis of the fracture mechanics deterministic argument is summarised in Figure 10. There are three major legs which input to the total argument.

(i) a proof test based assessment,

(ii) an assessment of reference defects,

(iii) detection of gas leakage from assumed defects of sub-critical size,

where a leg is defined as "an independent argument that in itself demonstrates safety". The argument for the Magnox reactor steel pressure vessels provides an example typical of the type of legs that can be assembled to formulate a multi-legged argument. It should be noted, however, that there are additional factors that, on this occasion, are accommodated, (a) the demonstration that the vessel operates at temperatures that are on the upper shelf of the fracture toughness curve, (b) a demonstration of the tolerance to large defects from the quality of vessel construction, and (c) plant monitoring.

(i) *Proof Test Assessment* In the proof-test assessment use is made of the fact that the vessels survived an original proof-test to over 1.5 times the design pressure to determine the maximum size of defects that could have been present in the welds. Growth of these defects is then assumed to occur by fatigue and creep

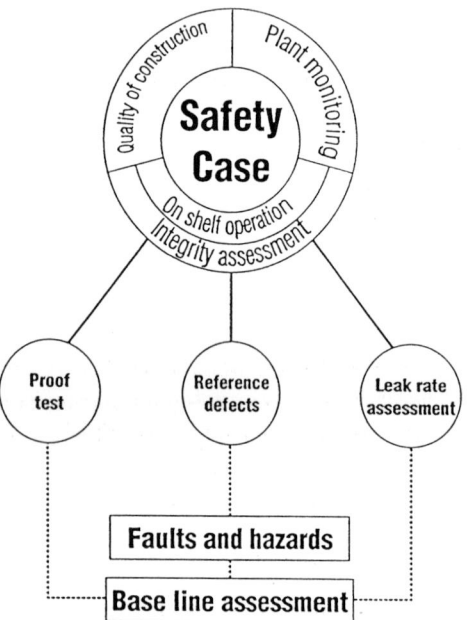

Fig. 10 Basis of the deterministic fracture mechanics based structural safety argument (case) for reactor steel pressure vessels showing the main legs of the baseline assessment; the additional factors, on-shelf operation, quality of construction and plant monotoring are also included.

processes in service and the extended defects are assessed to evaluate pressure margins against failure in future operation. This approach to determining possible defect sizes is grossly pessimistic since the derived defect depths are over 50% of the vessel wall thickness. For such pessimistic defects it is not always possible to demonstrate positive safety margins at all locations, in particular those subject to high in-service thermal or system stresses. However, in these cases, an argument can often be constructed which indicates that the length of weld at risk is small or the probability of a defect being in the depth range whereby it would survive the proof test yet fail in service is low.

(ii) *Reference Defects* As an alternative to the proof test approach reference defects 25mm deep are considered, including both fully extended defects and semi-elliptical defects with a 6:1 aspect ratio. The former is consistent with the requirements of ASME XI for pressure vessel assessments whilst the latter is considered to be a closer representation of the shape of defects that could be present in the vessels. It is judged very unlikely that defects in excess of 25mm deep would have been missed by the radiographic inspection carried out on all the welded joints following construction.

(iii) *Gas Leakage* The leak rate assessment lends additional support to the safety argument by providing confidence that should carbon dioxide leak from a postulated through-thickness crack in the pressure vessel, it would be readily detectable before the crack reached a critical size.

The overall logic of the approach adopted demonstrates that, in addition to positive pressure margins against failure for defects, consideration is given to the temperature margins which exist between the normal operating temperature of the vessel and the temperature at which the onset of upper shelf fracture toughness conditions occur. For this assessment the onset of upper shelf conditions is taken as the temperature at which there is only a 5% chance of brittle failure based upon the upper bound 95% confidence limit of the data. A positive temperature margin provides the added security that if the vessel were to fail it would do so in a progressive, ductile mode leading to carbon dioxide gas leakage.

The analysis route for the above assessment gives critical crack sizes for both part through and fully penetrating defects in the vessel. Further support for the safety argument is derived from a comparison of these critical sizes with the size of defects likely to be present in the welds judged from the quality of vessel construction and the results of ultrasonic inspection of welds in other component parts of the Magnox reactor pressure circuit.

3.3 Deterministic Structural Integrity Assessment

The deterministic method of assessment has been described in detail elsewhere Flewitt et al (1993), Flewitt et al (1994) and Wright (1994). Briefly, a fracture mechanics approach is adopted based on the R6 two criteria procedure, Milne et al (1986), referred to previously, Figure 8, for the assessment of structures containing defects. The main inputs to the fracture mechanics assessment are (i) depth, shape and position of a potential defect, (ii) materials properties, (iii) loading conditions, and (iv) geometry of the component. This procedure is used to determine the minimum failure pressure as a function of vessel temperature, defined as the pressure limit line, for different locations on the vessel. By comparing this failure line with the pressure/temperature limits set by the Operating

Rules, Figure 11, it is possible to derive temperature margins and reserve factors on pressure. Calculations have been carried out as a base line integrity assessment for a range of loading conditions including reactor start-up, normal steady state operation and a pressure fault loading condition defined by the setting of the safety relief valves. As an extension to this base line assessment, separate calculations have been carried out for a number of possible but unlikely fault and hazard loadings for the reference defect case only (see Figure 10).

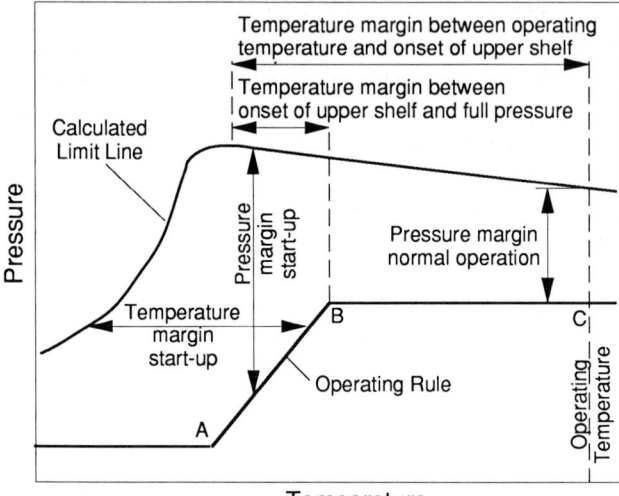

Fig. 11 A schematic diagram showing the pressure and temperature margins that can be derived when the Rule to which the plant is operated is compared with the calculated limit line.

Input Data

During service neutron irradiation degrades mechanical properties in the parent plate, but in particular the submerged arc weldments, English et al (1993). The compositions of the constituent materials for these Magnox reactor steel pressure vessels are given in Table 1. For these carbon manganese steels the neutron irradiation usually increases the brittle to ductile transition temperature, decreases the upper shelf energy and increases yield strength and hardness. There is also a corresponding reduction in overall ductility and work hardening rate. These arise from microstructural changes which include the formation of precipitates mainly copper and carbides, vacancy and interstitial clusters which produce small voids and dislocation loops and segregation of certain impurity elements to the ferrite grain boundaries. Thus neutron irradiation has the potential to enhance both diffusion and clustering mechanisms. Therefore it is important to have a detailed knowledge of the neutron fluence and irradiation temperature, combined with the composition of the material and its microstructure.

As a consequence the input data required for this deterministic fracture mechanics structural integrity assessment are:

(i) the start-of-life material properties for the submerged-arc weld metal including yield stress, ultimate tensile stress, fracture toughness, fatigue crack growth rates and creep properties including growth rates.

(ii) the temperature and neutron flux values at different locations in the pressure vessel,

(iii) the effect of neutron irradiation on the material properties,

(iv) the stresses arising from the loads that act in the vessel either during operation or an assumed fault condition.

Tensile properties at ambient temperature and Charpy impact energy values at -10°C were obtained during construction but no fracture toughness measurements were made. Charpy impact data over a wider temperature range have since been obtained from the monitoring scheme control specimens. To supplement the available data, tensile and fracture toughness tests have been undertaken over a wide range of temperatures on submerged-arc weld metal specimens obtained from material removed from the vessels during construction to accommodate the inlet and outlet ducts. The fracture toughness data have been analysed to provide brittle fracture and ductile crack initiation distributions at start-of-life, where initiation is taken as 0.2mm of stable crack extension. Survival statistics and the concept of competing risks have been used to analyse the data in the transition region where both modes of failure can occur, Moskovic (1993). Using this approach it is possible to estimate the start-of-life toughness distribution from the lower shelf, brittle fracture, to the upper shelf, ductile fracture and to determine the relative likelihood of brittle or ductile failure at any prescribed temperature. Figure 12 shows schematically the start-of-life fracture toughness distribution and the associated

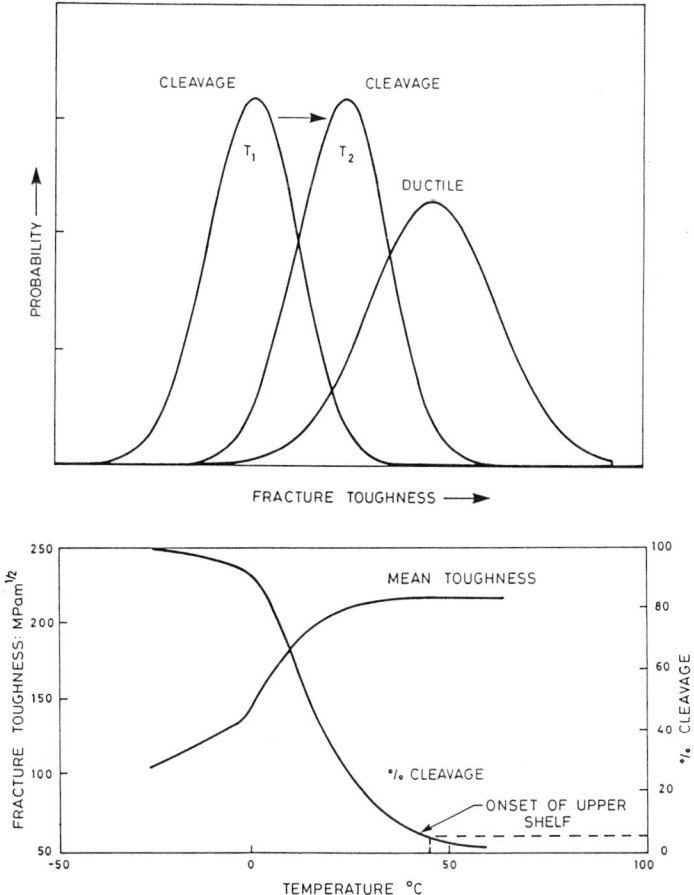

Fig. 12 *Derivation of a fracture toughness curve for submerged arc weld metal using competing risk analysis. (a) the interaction of cleavage and ductile probability distributions at two temperatures T1 and T2; (b) the variation of the mean fracture toughness with temperature and the corresponding calculated probability of cleavage fracture used for defining the onset of the upper shelf temperature (5% cleavage definition)*

probability of cleavage fracture plotted against temperature for submerged-arc weld metal. The temperature at which there is a 5% probability of cleavage fracture, at initiation; 0.2mm crack extension. This is equivalent to a 95% probability of ductile crack initiation and provides an appropriate statistical definition of the upper bound to the onset of upper shelf temperature. A small shift in transition toughness due to strain ageing is predicted to occur for submerged-arc weld metal and certain plate steels. This is included in the overall shift in transition toughness.

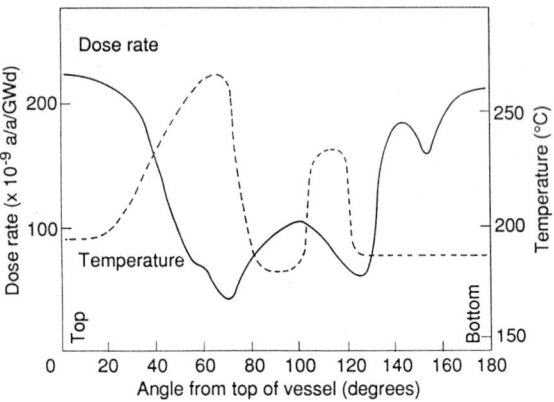

Fig. 13 A typical variation of neutron dose rate and temperature around a steel pressure vessel.

The neutron flux levels around the vessels have been derived from a combination of measurement and neutron transport calculations. Neutron activation measurements have been taken either on steel wires which were wrapped around the vessel or on samples of steel removed from redundant structures located within, but close to the steel pressure vessel. These measurements have provided flux levels in terms of the fission equivalent flux. However the effect of neutron damage on the material properties are more appropriately characterised in terms of displacements per atom (dpa) English et al (1993). Dpa estimates have been obtained by applying scaling factors, derived from complex neutron transport calculations, to the measured fission equivalent flux values. The models used to perform these calculations included the outer parts of the active core and the reflectors

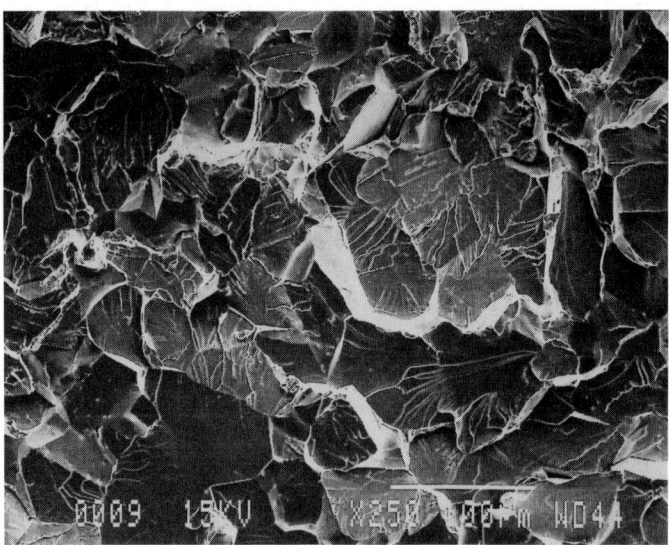

Fig. 14 Scanning electron fractograph showing a mixture of cleavage and intergranular fracture in a specimen fractured at a temperature in the brittle range.

in detail and representations of the core restraint, core support structure and other steelwork inside the vessels. The temperature at which a vessel operates is based on data collected from the permanent thermocouples which were attached to the outside surface of the vessel when it was constructed and on heat transfer models, the latter being used to interpolate between the discrete thermocouple positions on the vessel. The variation in the neutron dose rate and vessel temperature along a typical vessel meridian is shown in Figure 13.

The effect of neutron irradiation on the materials properties has been determined from changes measured on the surveillance scheme specimens. The data obtained from the separate reactor schemes have been analysed together to provide dose/damage trend curves. Upper shelf fracture toughness properties have been adjusted for the effects of irradiation on the basis of tests carried out on pre-cracked Charpy specimens. The toughness transition curve has been shifted upwards in temperature by an amount equal to the change in the Charpy 40J value. The surveillance scheme data have been supplemented with results from tests on specimens re-irradiated at an accelerated rate to higher dose levels. These demonstrate the changes in properties arising from copper precipitation, matrix damage and neutron induced phosphorus segregation to the grain boundaries which results in a proportion of intergranular embrittlement and fracture, Figure 14, in the submerged-arc weld metal. However, even when these weldments enter service there is a proportion of phosphorus segregated to the ferrite grain boundaries due to the original welding heat treatment cycle, Abbott et al, (1994). Although this level is insufficient to cause a significant amount of intergranular fracture, geometrical arguments demonstrate that a small proportion of intergranular fracture ≤2% will accompany the low temperature brittle cleavage fracture, Abbott et al, (1994). Dose/damage trend curves which accommodate all three embrittlement mechanisms have been derived for use in predicting the shift in transition temperature, Bolton et al (1994), and a typical example is given in Figure 15.

The best estimate irradiated fracture toughness at any location in the vessel can be determined by using the appropriate dose and dose/damage law to calculate the shift in brittle toughness and change in upper shelf toughness and then adding these changes to the start-of-life properties. This

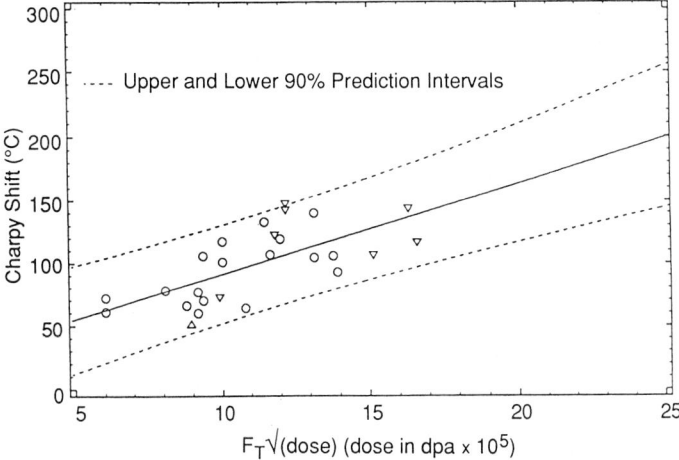

Fig. 15 Variation of the Charpy impact energy, 40J, temperature shift for submerged arc weld metal with neutron dose in displacements per atom (dpa) at an exposure temperature of 190 °C

process becomes more complicated when the uncertainties in the start-of-life toughness, dose and dose/damage law are taken into account. Hence a computer program is used which determines the irradiated cleavage and ductile initiation toughness distributions by combining these three distributions. This program determines the combined irradiated toughness/temperature distribution using a competing risks approach described by Moskovic (1993) and the temperature at which there is a 95% probability of ductile crack initiation. Figure 16 shows the predicted change in fracture toughness and cleavage fracture probability for a typical submerged-arc weld location after thirty years service. The temperature at which there is a 95% probability of ductile crack initiation is used when specifying the Operating Rules for a reactor pressure vessel to ensure a ductile condition when significantly stressed.

For conservatism the deterministic fracture mechanics assessment is based on the most embrittled inside surface material properties, even though the crack tip of the reference flaw is at a depth of ≥25mm beneath the inside surface. Since the overall fracture behaviour will be governed by the material properties in the remaining ligament, it follows that a more realistic assessment would be obtained by using the initiation toughness at the crack tip which would provide increased safety margins during start-up and increased temperature margins above the 95% ductile temperature during normal operation. To further strengthen the present position with regard to the materials properties for submerged-arc weld metal, there is a continuing supporting work programme. Further specimens will be withdrawn from the monitoring scheme together with accelerated irradiation tests to add to the dose/damage database. However, an ambitious plan has been set in place which will remove samples of submerged-arc weld metal from a shutdown reactor pressure vessel in 1995/1996. These samples, 160mm dia, will be trepanned from a location on the vessel wall where neutron doses on the inside surface

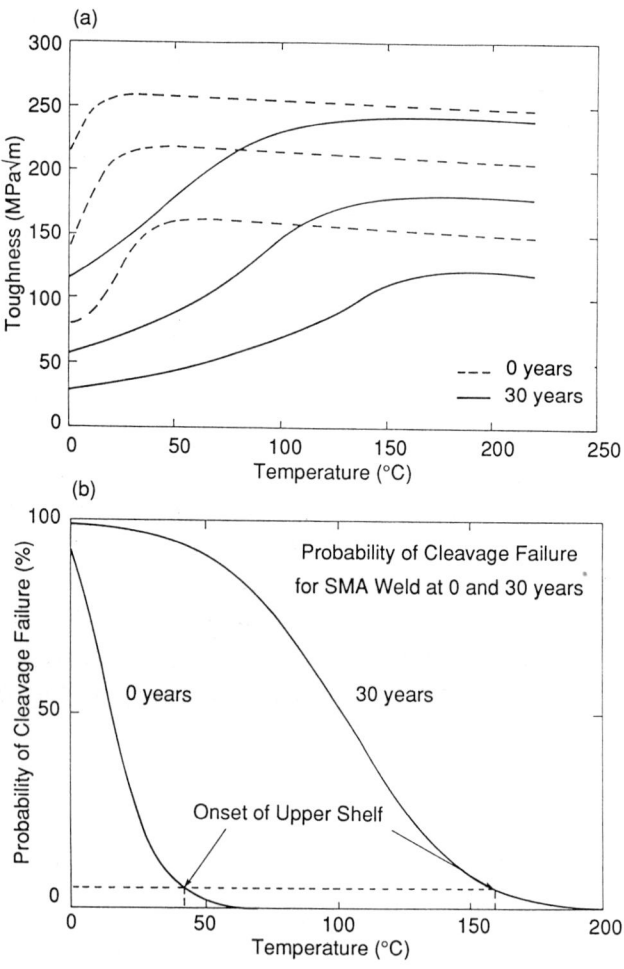

Fig. 16 Onset of upper shelf for submerged arc weld metal showing (a) the best estimate and upper and lower bounds for the fracture toughness at start of life and after 30 years service (b) the corresponding calculated probability of cleavage failure defining the onset of the upper shelf temperature.

will be in excess of the highest dose received by the remaining operating Magnox reactor steel pressure vessels. The neutron flux is attenuated when passing through the vessel wall and Charpy impact test specimens will be obtained from different depths in these samples. This will allow the data to be compared with the predictions from the relevant dose/damage trend curve. The results from this supporting work will add to the present understanding of the embrittlement process and should improve prediction and alleviate conservatisms used in the deterministic assessment of the operating pressure vessels.

During service stresses occur in a steel pressure vessel from various types of loading including internal pressure, and thermal and system loading. Although the vessels were subject to a stress relief heat treatment a low level of residual stress is also assumed to exist in the welds. In the base line assessments two pressure levels have been considered, the first equivalent to the normal operating pressure, and the second relating to the safety relief valve settings. Thermal stresses arise in regions of the vessel where there are large temperature gradients, e.g. across the gas seal or where the vessel insulation is discontinuous. System loads are applied to the vessel via the gas ducts and result from differential displacement between the vessels and the boiler shells. Stresses from these various types of loading have been obtained by finite element analysis.

In recent assessments the implications of loads occurring as a result of a range of different faults and hazards have been considered assuming reference size defects to be present in the vessels, Figure 10. These include seismic loading, dropped loads, boiler tube failure and anomalies in the duct hanger system. The principle adopted in examining the effects of such loads on the integrity of the vessels is to identify a set of bounding faults and hazards representing the most severe events which could conceivably occur over their lifetime. Less frequent but more severe events are not considered on the basis of the low probability of large defects existing together with such remote loadings. This approach is broadly similar to that used in ASME III for pressurised water reactor components where infrequent events are considered separately by applying different allowable stresses or service limits compared with normal operation.

At the time the pressure vessels were designed it was believed that no non-destructive inspection would be possible once the reactors entered service. Therefore considerable reliance was placed on the inspections carried out during and after vessel construction to ensure the quality of manufacture. The principal inspection technique available at that time was radiography using X-rays and this was used to inspect all the butt and seam welds in the vessel. In later vessels a limited amount of ultrasonic inspection was also employed. Where any defects were found outside a stringent acceptance code the welds were ground to remove the defects and re-welded. A final check on the quality of construction was made through a proof pressure test to over 1.5 times the design pressure plus a leak test. Large areas of the vessels were subsequently covered with internal and external insulation and problems of direct access plus the high radiation levels that exist, even with the reactors shut down, made any further inspection impossible for many years into the operating life of the vessels.

However, due to the recent technical advances in remote inspection techniques, re-inspection of some areas of the vessels has become feasible and an ultrasonic inspection has been carried out on a limited length of weld on the vessels at two power stations. At one power station the vessel is not insulated internally and this has allowed access to the welds in the region of the outlet gas duct nozzles from inside the vessel, Bowring et al (1993). An ultrasonic probe head was carried on a rotating beam at the end of a multi-link manipulator which was deployed down a refuelling standpipe. The system allowed line scans to be taken across the weld and with the beam rotated in increments through a full 360° the entire length of the nozzle attachment weld could be inspected. An alternative approach where access to the welds from inside the vessel is prevented by the internal insulation was carried out from outside the vessel, with the ultrasonic probe carried on a remotely controlled trolley, Curry and Burrows (1993). The total length of weld inspected at the two power stations was in excess of 75 metres: no defects of any significant depth were detected. These inspection results provide confidence in the high quality of construction of the Magnox reactor pressure vessels.

3.4. Consequences for Pressure Vessels and Operational Procedures

The purpose of the integrity assessment is to demonstrate that adequate pressure reserve factors and temperature margins exist over the full operating range of the vessels, and that should a fully penetrating defect be present in the vessel, the leakage of carbon dioxide coolant gas would be detected well before the defect reaches a critical length. The strength of the overall safety argument depends on the different approaches adopted whereby any potential weakness in one leg of the case is offset by the strength of the remaining legs.

To strengthen the argument it has been appropriate to make a number of changes to the operating procedures and to carry out modifications to the plant by providing additional measuring/ detection equipment. The integrity assessments have indicated that after taking account of the various uncertainties in the prediction method, the upward shift in the fracture toughness transition curve is typically of the order 150°C. To accommodate this change in the transition temperature, changes have been made progressively to the Operating Rules which set the pressure/temperature limits during a reactor start-up, Figure 17, and to ensure that the vessels are not pressurised fully until ductile fracture conditions have been achieved in the weld metal. In addition, it has been necessary to raise steady state vessel operating temperatures to ensure that a sufficiently large margin exists above the temperature at which the onset of upper shelf conditions occur. Generally this has been achieved by reducing the shield cooling air flow to the external surface of the vessel. However, in these circumstances, care has to be taken so that any temperature limits on the concrete biological shield or the shield cooling fans are not exceeded. Alternatively, the temperature of the bottom dome region of the vessel has been increased by increasing reactor gas inlet temperature but with a consequent reduction in reactor power output.

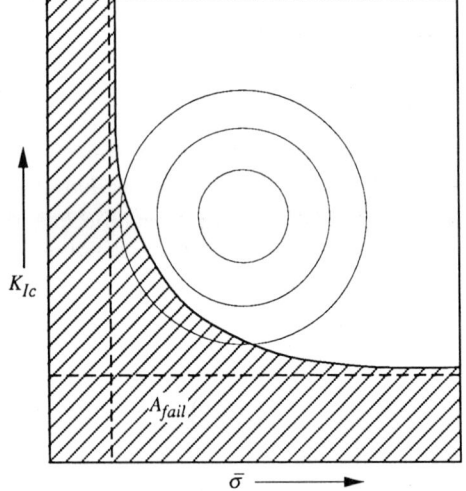

Fig. 17 Schematic diagram showing a failure assessment diagram and integration over material property distributions; material property contours are shown after transformation to K_{IC} and space for a given defect size.

The inclusion of the leak rate assessment as part of the overall safety argument has required improvements to be made to the installed carbon dioxide gas leak detection systems. These were intended originally to provide a general indication of any gas leakage problems and are not capable of giving accurate measurements of carbon dioxide leak rate at the levels predicted in the assessments. New detection systems have been installed which measure the increase in carbon dioxide gas concentration in the shield cooling air and the air flow and the air flow rate thus providing a detection capability down to about 1 tonne/day. In addition, to cover those locations where high thermally induced bending stresses cause crack closure and hence limit gas leakage, leak testing has been initiated during reactor start-up when isothermal conditions still exist.

4. Concluding Comments

Deterministic structural integrity arguments are formulated and used to ensure the safe operation of a variety of high integrity structures and components, for example nuclear power electrical generating plant. These arguments are based upon a knowledge of the plant design and operating requirements combined with the inspection, diagnosis and assessment of any existing or hypothetical defects. This approach to structural integrity assessments leads to an overall argument (or Safety Case) that is often multi-legged, where each leg is formulated using **conservative** assumptions with respect to the size and distribution of potential defects and input parameters. However the strength of the overall argument is judged from an evaluation of the relative strength and weakness of each leg. Such an approach is demonstrated to be reasonable by the example described in Section 3 for Safety Cases developed for the four Magnox reactor steel pressure vessel power stations operated by Nuclear Electric plc in the UK. In this example of a deterministic fracture mechanics based structural integrity argument, it is clear that the procedure adopted provides a secure and robust basis for setting Operating Rules which ensure the continued safe operation of this plant. The oldest of these power stations has now operated successfully for over 30 years and work is in hand to justify continued operation up to 40 years for all of the reactors. The degradation of the vessel steels due to neutron irradiation and thermal ageing effects will be taken into account and revisions proposed to the Operating Rules to preserve adequate pressure and temperature margins.

However it is clear that as an example of a typical deterministic argument for a high integrity structure or component undue conservatisms have to be introduced which could be relieved if there were a better understanding of the materials behaviour and adequate accommodation of materials variables via an improved mechanistic understanding that would allow increased confidence in the prediction of the overall operating life of the plant or components. In general, there is certainly a need to improve the ability to predict the performance of a material from a knowledge of its composition, microstructure and processing history. Although there is an increased recognition of the need for plant surveillance specimens to assist in the assessment and underwrite of the integrity of plant and components, there remains the need to improve both accelerated life testing and the associated methodology. This means that accelerated life testing for a range of temperatures, environments and stress states have to be combined with predictive methodology and developed beyond the present understanding. It is certainly not satisfactory when assessing high integrity structures and components to simply rank materials behaviour under accelerated and, therefore, artificial test conditions. Rather there is a need to develop assessment methodology and mechanical property data which when combined with a mechanistic model will allow a quantitative life assessment with the appropriate level of confidence to assure reliability and ensure failure will not occur. Until this stage is achieved to the necessary level, high cost conservatisms have to be built into the structural integrity arguments and these may have to be supported by high cost, non-destructive inspection and monitoring of the operating conditions combined with the use of data obtained from associated surveillance schemes. An alternative is to consider a probabilistic structural integrity argument in total or as a part input to the deterministic argument.

Detailed probabilistic fracture mechanics methodology based on the well established R6 procedure for the assessment of structures and components containing defects are at an early stage of development, Gates et al (1990), Francis (1993). Two computer programs are used to implement this methodology (a) the STAR6 program uses analytical and numerical techniques to perform the required integrations whereas (b) the PROF code uses Monte Carlo simulation methods, Wilson (1990). However, in principle a probability density function is required for each input quantity. As a consequence, the probability of failure, P_f, in each case takes the form

$$P_f = \int_0^w f(a) \iint_{A_{fail}} P_1(K_{Ic}) P_2(\overline{\sigma}) \, d\overline{\sigma} \, dK_{Ic} \, da$$

where $f(a)$, $P_1(K_{Ic})$ and $P_2(\overline{\sigma})$ are the probability density functions for a defect of size a, the fracture toughness, K_{Ic}, and the flow stress, $\overline{\sigma}$, of the material. The limit of integration, w, is the component section thickness and the integration region A_{fail} in Figure 17 is related to the failure region of the R6 diagram shown in Figure 8. In practice, due to computing limitations only the defect through-wall extent, fracture toughness and flow stress distributions are used in STAR6 while deterministic values are used for all other quantities such as dimensions and stresses. PROF can consider up to twelve probabilistic variables but requires much greater computing times compared with STAR6 to give statistically reliable results. For example, this approach could be applied to the assessment described in Section 3 for steel pressure vessel. However to apply the probabilistic methodology defined by the above equation, it would be necessary to take account of factors such as fatigue crack growth, stable tearing, critical crack length, fault loadings, pre-service proof tests and competing ductile and cleavage fracture mechanisms. The incorporation of these into the probabilistic argument would add considerably to the complexity of the actual expressions to be evaluated but the basic principles remain. At present it is more appropriate that probabilistic methods should be considered to enhance, rather than replace, the traditional deterministic approaches to structural assessments. However they do offer a realistic way forward for integrity assessments to accommodate the variability of all the input parameters including those for the material mechanical properties which is a limitation and conservatism for deterministic arguments.

Acknowledgement

The author would like to acknowledge the significant efforts of many colleagues within Nuclear Electric plc who have contributed to the development of the assessment methodology, the collection of relevant input data and assisted in formulating my ideas presented in this paper. The paper is published with permission of the Director of Engineering, Nuclear Electric plc.

References

Abbott K, Moskovic R, and Flewitt P E J (1994). A Comment on Intergranular Fracture in the Cleavage Temperature Range. *J. Mat. Sci. and Tech.*, **10**, 813.

Abbott K, Moskovic R, Wild R K and Flewitt P E J (1994). Low Temperature Fracture of C-Mn Submerged-Arc Weld Metal. *J. Mat. Sci. and Tech.* (in press).

Cocks A C F and Ashby M F (1982). *Progress in Materials Science*, **27**, 189.

Bowring N A, Rule J R, Smith A L, Goodhind G L and Meredith P J (1993). Ultrasonic Inspection of a Magnox Reactor Pressure Vessel from the Inside. *BNES Conference, Remote Techniques for Nuclear Plant*, Paper No. 14, Nuclear Energy Society (London), pp. 90-98.

Bolton C J, Buswell J T, Jones R B, Moskovic R, Priest R H (1995), The Modelling of Irradiation Embrittlement in Submerged-Arc Welds, Effects of Radiation on Materials: *17th International Symposium, American Society for Testing and Materials*, Philadelphia, to be published.

Curry A and Burrows M S (1993). Inspection of a Magnox Reactor Pressure Vessel from the Outside. *BNES Conference, Remote Techniques for Nuclear Plant*, Paper No. 15, Nuclear Energy Society (London), 99-103.

Doig P and Gasper B C (1992), An Overview of Plant Structural Integrity Assessment. In: *Structural Integrity Assessment* (ed. P Stanley), Elsevier Applied Science (London) p. 163.

Endo T (1994), Progress on Life Assessment and Design Methodology for Fossil Power Plant Components, *Int. J. Press. Vess. and Piping* **57**, 7.

English C A, Fudge A H, McElroy R J, Phythian W J, Buswell J T, Bolton C J and Jones R B (1993), Approach and Methodology of Condition Assessment of Thermal Reactor Pressure Vessels. *Int. J. Pres. Vess. and Piping*, **54**, 49-88.

Flewitt P E J and Williams G H (1994), Structural Integrity Safety Arguments for Nuclear Generating Plant, In: *Engineering Integrity Assessment* (eds. Edwards J H, Kern J and Stanley P), EMAS (London), p. 323.

Flewitt P E J, Williams G H and Wright M B (1993), Integrity of Magnox Reactor Steel Pressure Vessels. *J. Mat. Sci. and Tech.* **9**, 75-82.

Flewitt P E J, Williams G H and Graham W J (1994), The Structural Integrity Assurance of Magnox Reactor Steel Pressure Vessels. In: *The Inspection and Structural Validation of Nuclear Power Plant*, I. Mech. E (London), p. 11.

Francis A (1993), *A Probabilistic Fracture Mechanics Methodology Allowing for a Different Failure Mechanism at the Final Loading to that at the Initial Loading*, Nuclear Electric Report TIGN/REP/0002/93.

Gates R S, Francis A, Kolbuszewski M, Wilson R and Windle P L (1990), *A Method for Calculating the Probability of Failure of Structures using Elastic-Plastic Fracture Mechanics*, Nuclear Electric Report, TD/SID/REP/0030.

Grey (1972), Investigations into the Consequences of the Failure of a Turbine Generator at Hinkley Point 'A' Power Station, *Proc. Inst. Mech. Eng.* **186**, 399.

Holliday W C and Noone M J (1961), *Metallurgical Problems Associated with the Construction of Gas Cooled Reactor Pressure Vessels. Steels for Reactor Pressure Circuits*, Iron and Steel Institution, Special Report No. 69, pp. 207-239.

Kalderon, D (1972), Steam Turbine Failure at Hinkley Point 'A', *Proc. Inst. Mech. Eng.* **186**, 341.

Milne I and Ainsworth R A, Dowling A R and Stewart A T (1986), *Assessment of the Integrity of Structures Containing Defects*, CEGB Report R/H/R6 Rev. 3.

Moskovic R, (1993), Statistical Analysis of Censored Fracture Toughness Data in the Ductile to Brittle Transition Temperature Region. *Eng. Fract. Mech.*, **44**, 21-41.

Neubauer B and Bietenbeck F (1989), The Potential of Medium and Long Term Life Predictions for High Temperature Components in Power Plants, *Int. J. Pres. Vess. and Piping*, **39**, 57.

H M Nuclear Installations Inspectorate (1992), *Safety Assessment Principles for Nuclear Plant* HM Nuclear Installations Inspectorate, October (HMSO London).

Poynor J F (1969), Design and Construction of the Sizewell Reactor Pressure Vessel. Performance of Nuclear Power Components, In: *Proceedings of Symposium held in Prague*, pp. 379-388.

Prager M (1991), Assuring Reliability of Advanced Materials in Severe Applications, *Eng. Frac. Mech.*, **40**, 921.

Swift T (1994), Damage Tolerance Capability, *Fatigue*, **16**, 75.

Wannenberg J, Klintworth G C and Rauth A D (1992), *Int. J. Pres. Vess. and Piping*, **50**, 255.

Wilson R (1990), *User's Guide to the Probabilistic Fracture Mechanics Computer Code STAR6,* Nuclear Electric Memorandum TD/SID/MEM/0007.

Wright M B (1994), Magnox Reactor Steel Pressure Vessels, Fracture Mechanics Based Assessment Procedure, In: *Proceedings Saclay International Seminar on Structural Integrity*, 28-29 April 1994 (in press).